Nanoparticles for Catalysis

Special Issue Editors

Hermenegildo García

Sergio Navalón

MDPI • Basel • Beijing • Wuhan • Barcelona • Belgrade

Special Issue Editors
Hermenegildo García
Universidad Politécnica de Valencia
Spain

Sergio Navalón
Universidad Politécnica de Valencia
Spain

Editorial Office
MDPI AG
St. Alban-Anlage 66
Basel, Switzerland

This edition is a reprint of the Special Issue published online in the open access journal *Nanomaterials* (ISSN 2079-4991) from 2015–2016 (available at: http://www.mdpi.com/journal/nanomaterials/special_issues/nano_catal).

For citation purposes, cite each article independently as indicated on the article page online and as indicated below:

Author 1; Author 2. Article title. *Journal Name* **Year**, *Article number*, page range.

First Edition 2017

ISBN 978-3-03842-536-6 (Pbk)
ISBN 978-3-03842-537-3 (PDF)

Table of Contents

About the Special Issue Editors v

Preface to "Nanoparticles for Catalysis" vii

Jose M. Palomo and Marco Filice
Biosynthesis of Metal Nanoparticles: Novel Efficient Heterogeneous Nanocatalysts
Reprinted from: *Nanomaterials* **2016**, *6*(5), 84; doi: 10.3390/nano6050084 1

Nimesh Shah, Pallabita Basu, Praveen Prakash, Simon Donck, Edmond Gravel, Irishi N. N. Namboothiri and Eric Doris
Supramolecular Assembly of Gold Nanoparticles on Carbon Nanotubes: Application to the Catalytic Oxidation of Hydroxylamines
Reprinted from: *Nanomaterials* **2016**, *6*(3), 37; doi: 10.3390/nano6030037 17

Muhammad Humayun, Zhijun Li, Liqun Sun, Xuliang Zhang, Fazal Raziq, Amir Zada, Yang Qu and Liqiang Jing
Coupling of Nanocrystalline Anatase TiO_2 to Porous Nanosized $LaFeO_3$ for Efficient Visible-Light Photocatalytic Degradation of Pollutants
Reprinted from: *Nanomaterials* **2016**, *6*(1), 22; doi: 10.3390/nano6010022 25

Weiyi Ouyang, Ewelina Kuna, Alfonso Yepez, Alina M. Balu, Antonio A. Romero, Juan Carlos Colmenares and Rafael Luque
Mechanochemical Synthesis of TiO_2 Nanocomposites as Photocatalysts for Benzyl Alcohol Photo-Oxidation
Reprinted from: *Nanomaterials* **2016**, *6*(5), 93; doi: 10.3390/nano6050093 35

Karen Leus, Jolien Dendooven, Norini Tahir, Ranjith K. Ramachandran, Maria Meledina, Stuart Turner, Gustaaf Van Tendeloo, Jan L. Goeman, Johan Van der Eycken, Christophe Detavernier and Pascal Van Der Voort
Atomic Layer Deposition of Pt Nanoparticles within the Cages of MIL-101: A Mild and Recyclable Hydrogenation Catalyst
Reprinted from: *Nanomaterials* **2016**, *6*(3), 45; doi: 10.3390/nano6030045 47

Anurag Kumar, Pawan Kumar, Chetan Joshi, Manvi Manchanda, Rabah Boukherroub and Suman L. Jain
Nickel Decorated on Phosphorous-Doped Carbon Nitride as an Efficient Photocatalyst for Reduction of Nitrobenzenes
Reprinted from: *Nanomaterials* **2016**, *6*(4), 59; doi: 10.3390/nano6040059 58

Dimitrios Andreou, Domna Iordanidou, Ioannis Tamiolakis, Gerasimos S. Armatas and Ioannis N. Lykakis
Reduction of Nitroarenes into Aryl Amines and N-Aryl hydroxylamines via Activation of $NaBH_4$ and Ammonia-Borane Complexes by Ag/TiO_2 Catalyst
Reprinted from: *Nanomaterials* **2016**, *6*(3), 54; doi: 10.3390/nano6030054 72

Wenjing Zhang, Yin Cai, Rui Qian, Bo Zhao and Peizhi Zhu
Synthesis of Ball-Like Ag Nanorod Aggregates for Surface-Enhanced Raman Scattering and Catalytic Reduction
Reprinted from: *Nanomaterials* **2016**, *6*(6), 99; doi: 10.3390/nano6060099 84

Huishan Shang, Kecheng Pan, Lu Zhang, Bing Zhang and Xu Xiang
Enhanced Activity of Supported Ni Catalysts Promoted by Pt for Rapid Reduction of Aromatic Nitro Compounds
Reprinted from: *Nanomaterials* **2016**, *6*(6), 103; doi: 10.3390/nano6060103 95

Wenhai Ji, Weihong Qi, Shasha Tang, Hongcheng Peng and Siqi Li
Hydrothermal Synthesis of Ultrasmall Pt Nanoparticles as Highly Active Electrocatalysts for Methanol Oxidation
Reprinted from: *Nanomaterials* **2015**, *5*(4), 2203–2211; doi: 10.3390/nano5042203 109

Xiaohua Wang, Yueming Li, Shimin Liu and Long Zhang
N-doped TiO_2 Nanotubes as an Effective Additive to Improve the Catalytic Capability of Methanol Oxidation for Pt/Graphene Nanocomposites
Reprinted from: *Nanomaterials* **2016**, *6*(3), 40; doi: 10.3390/nano6030040 116

Chih-Chun Chin, Hong-Kai Yang and Jenn-Shing Chen
Investigation of MnO_2 and Ordered Mesoporous Carbon Composites as Electrocatalysts for Li-O_2 Battery Applications
Reprinted from: *Nanomaterials* **2016**, *6*(1), 21; doi: 10.3390/nano6010021 125

Alessandro Minguzzi, Gianluca Longoni, Giuseppe Cappelletti, Eleonora Pargoletti, Chiara Di Bari, Cristina Locatelli, Marcello Marelli, Sandra Rondinini and Alberto Vertova
The Influence of Carbonaceous Matrices and Electrocatalytic MnO_2 Nanopowders on Lithium-Air Battery Performances
Reprinted from: *Nanomaterials* **2016**, *6*(1), 10; doi: 10.3390/nano6010010 138

About the Special Issue Editors

Hermenegildo García is Full Professor at the Technical University of Valencia and staff of the Instituto de Tecnología Química, a joint center of the Technical University of Valencia and the Spanish National Research Council. Prof. Garcia has been active in the field of heterogeneous catalysis working with zeolites, mesoporous materials, metal organic frameworks, graphene and nanoparticles, particularly supported gold nanoparticles. He has published over 600 papers and has filed over 25 patents, two of them in industrial exploitation. Prof. Garcia is the 2016 Rey D. Jaime I award in New technologies, Doctor Honoris Causa from the University of Bucharest and the recipient of the 2011 Janssen-Cilag award given by the Spanish Royal Society of Chemistry and the 2008 Alpha Gold of the Spanish Society of Glass and Ceramics.

Sergio Navalón is Assitant Professor at the Department of Chemistry of the Technical University of Valencia (Valencia, Spain). He graduated in Chemical Engineering in 2003 and obtained his PhD in 2010 at the Technical University of Valencia (UPV). His research focuses on the development of heterogeneous (photo)catalysts based on graphene, porous materials and nanoparticles, as well as green chemistry processes. He is the author of about 50 publications, 2 chapter books and co-editor of one book.

Preface to "Nanoparticles for Catalysis"

Nanoscience emerged in the last decades of the 20th century with the general aim to determine those properties that appear when small particles of nanometric dimensions are prepared and stabilized. One of the clearest examples of specific properties of nanoparticles (NPs) is the ability to catalyze reactions by interacting with substrates and reagents. These unique properties of NPs as catalysts derive from the large percentage of coordinatively unsaturated atoms located at the surface, edges and corners of the NPs compared to the total number of atoms. Particularly those atoms located at steps, corners and edges of NPs exhibit the highest catalytic activity due to their low coordination number and their high tendency to increase this number by coordinating with substrates and other species in the surroundings.

While NPs can be prepared by different procedures, their main drawback is their tendency to undergo agglomeration as they increase in size, thereby reducing the energy associated with large surface area. One general methodology to circumvent this problem is by adsorbing these NPs on large surface area of insoluble solids that by means of strong interactions are able to adsorb NPs on their surface, stabilizing them against sintering and growth. In addition to this role, supports can also play an additional role in the catalysis providing acidity/basicity or by tuning the electronic density of the NPs.

The present Special Issue of Nanomaterials falls with the domain of NPs applied to catalysis. In their review, Palomo and Filice illustrate the synthesis of metal NPs by reduction of the corresponding metal precursors using biomolecules, plant extracts and even microorganisms [1]. The general advantage of these methods to form metal NPs is that no chemical reducing agents are employed and, in this sense, the synthesis based on the use of biomolecules can be environmentally more benign and sustainable, resulting in the generation of fewer chemical residues. In addition, biomolecules can also act as ligands of the metal NPs and, in this way, they not only reduce the metal ions to the metallic state, but also contribute to the formation of stable suspensions in water or other green solvents.

As discussed above, the solid support adsorbing NPs should primarily increase the stability of otherwise highly reactiveNPs. In their study, Doris and co-workers have used carbon nanotubes (CNTs) as scaffolds to deposit a particular organic polymer on which gold NPs can be deposited [2]. Specifically, CNTs were covered by the polymer resulting from light-promoted nitrilotriacetic-diyne polymerization followed by subsequent Au deposition. The resulting construct has a defined 1D morphology imparted by the CNT scaffold and was used to catalyze the aerobic oxidation of hydroxylamines to nitrones. Besides thermal oxidations, NPs exhibit interesting properties as photocatalysts. This aspect has also been covered in this Special Issue with two contributions showing the activity of small TiO_2 for the removal of pollutants [3] and for the selective oxidation of benzylalcohol to benzaldehyde [4].

In addition to oxidations, NPs, particularly metal NPs, have also general catalytic activity in reductions. In one of the articles of this Special Issue, van der Voort and co-workers have encapsulated Pt NPs inside the cages of a metal-organic framework (MIL-101-Cr) by atomic layer deposition [5]. Due to their internal porosity and crystalline structure, metal-organic frameworks offer confined spaces with regular dimensions in which NPs can be stabilized by geometrical constrains. Besides molecular hydrogen, other reducing agents such as hydrazine [6], $NaBH_4$ [7–9] and aminoborane [7] can also be activated by metal NPs to promote the reduction of different compounds. Nitroaromatics are among the preferred model substrates when the purpose is to determine the activity of catalysts, due to the possibility to follow the course of the reaction by absorption spectroscopic [7–9]. As in the case of oxidations, reductions can also be promoted by light. In one of these examples, Jain and co-workers have used a phosphorous-doped carbon nitride semiconductor that has been modified by Ni NPs and this system is able to promote light-assisted nitroarene to aniline reductions by using hydrazine as reducing agent [6].

Supported metal NPs are also suitable for use in the promotion of electrocatalytic reactions. Due to their interest in fuel cells, Qi and co-workers have prepared ultra-small size Pt NPs stabilized by polyvinylpyrrolidone that have been used as electrodes in methanol oxidation to CO_2 [10]. This reaction is of great importance for low temperature fuel cells. In a related work, Li and co-workers have shown that N-doped TiO_2 nanotubes as additives also increase the activity of graphene supported Pt composites in electrocatalytic methanol oxidation [11].

Continuing in the domain of renewable energies, NPs also offer considerable advantages with respect to other materials for the development of more efficient and cyclable electrodes for batteries. One type of battery that is very promising in the future due to high energy density and availability is the Li-O_2 battery. One of the main limitations of this type of batteries is the development of an efficient catalyst for the electrochemical reactions involving gas-phase O_2 and solid lithium oxides. In this context, Chen and co-workers have shown that MnO_2 NPs supported on carbon composites exhibit a high catalytic activity for oxygen reduction reaction [12], one of the key reactions involved. Similarly, MnO_2-doped with Ag dispersed in porous carbonaceous matrices are suitable gas diffusion electrodes for this type of batteries [13].

Overall, the present Special Issue shows the breath of applications and the potential of NPs in various fields going from thermal catalysis of liquid and gas phase reactions to photocatalysis using visible or solar light and electrocatalysis. In all these fields, the activities of nanomaterials have been found to exceed those of other types of particles. The target in this area is to further reduce the particle size, while gaining control of the morphology and facet orientation of the NPs, as well as their stabilization, without diminishing their activity. Another general challenge is to replace noble and critical metals by other abundant base transition metals. These targets will surely be accomplished in the near future.

Conflicts of Interest: The authors declare no conflict of interest.

Hermenegildo García and Sergio Navalón

Special Issue Editors

References

1. Palomo, J.M.; Filice, M. Biosynthesis of metal nanoparticles: novel efficient heterogeneous nanocatalysts. *Nanomaterials* **2016**, *6*, 84.
2. Shah, N.; Basu, P.; Prakash, P.; Donck, S.; Gravel, E.; Namboothiri, I.N.N.; Doris, E. Supramolecular assembly of gold nanoparticles on carbon nanotubes: Application to the catalytic oxidation of hydroxylamines. *Nanomaterials* **2016**, *6*, 37.
3. Humayun, M.; Li, Z.; Sun, L.; Zhang, X.; Raziq, F.; Zada, A.; Qu, Y.; Jing, L. Coupling of nanocrystalline anatase TiO_2 to porous nanosized $LaFeO_3$ for efficient visible-light photocatalytic degradation of pollutants. *Nanomaterials* **2016**, *6*, 22.
4. Ouyang, W.; Kuna, E.; Yepez, A.; Balu, A.M.; Romero, A.A.; Colmenares, J.C.; Luque, R. Mechanochemical synthesis of TiO_2 nanocomposites as photocatalysts for benzyl alcohol photo-oxidation. *Nanomaterials* **2016**, *6*, 93.
5. Leus, K.; Dendooven, J.; Tahir, N.; Ramachandran, R.K.; Meledina, M.; Turner, S.; van Tendeloo, G.V.; Goeman, J.L.; van der Eycken, J.V.; Detavernier, C.; et al. Atomic layer deposition of Pt nanoparticles within the cages of MIL-101: A mild and recyclable hydrogenation catalyst. *Nanomaterials* **2016**, *6*, 45.
6. Kumar, A.; Kumar, P.; Joshi, C.; Manchanda, M.; Boukherroub, R.; Jain, S.L. Nickel decorated on phosphorous-doped carbon nitride as an efficient photocatalyst for reduction of nitrobenzenes. *Nanomaterials* **2016**, *6*, 59.
7. Andreou, D.; Iordanidou, D.; Tamiolakis, I.; Armatas, G.S.; Lykakis, I.N. Reduction of nitroarenes into aryl amines and *N*-aryl hydroxylamines via activation of $NaBH_4$ and ammonia-borane complexes by Ag/TiO_2 catalyst. *Nanomaterials* **2016**, *6*, 54.
8. Zhang, W.; Cai, Y.; Qian, R.; Zhao, B.; Zhu, P. Synthesis of ball-like Ag nanorod aggregates for surface-enhanced Raman scattering and catalytic reduction. *Nanomaterials* **2016**, *6*, 99.
9. Shang, H.; Pan, K.; Zhang, L.; Zhang, B.; Xiang, X. Enhanced activity of supported Ni catalysts promoted by Pt for rapid reduction of aromatic nitro compounds. *Nanomaterials* **2016**, *6*, 103.

10. Ji, W.; Qi, W.; Tang, S.; Peng, H.; Li, S. Hydrothermal synthesis of ultrasmall Pt nanoparticles as highly active electrocatalysts for methanol oxidation. *Nanomaterials* **2015**, *5*, 2203–2211.
11. Wang, X.; Li, Y.; Liu, S.; Zhang, L. N-doped TiO_2 nanotubes as an effective additive to improve the catalytic capability of methanol oxidation for Pt/graphene nanocomposites. *Nanomaterials* **2016**, *6*, 40.
12. Chin, C.-C.; Yang, H.-K.; Chen, J.-S. Investigation of MnO_2 and ordered mesoporous carbon composites as electrocatalysts for Li-O_2 battery applications. *Nanomaterials* **2016**, *6*, 21.
13. Minguzzi, A.; Longoni, G.; Cappelletti, G.; Pargoletti, E.; di Bari, C.; Locatelli, C.; Marelli, M.; Rondinini, S.; Vertova, A. The influence of carbonaceous matrices and electrocatalytic MnO_2 nanopowders on lithium-air battery performances. *Nanomaterials* **2016**, *6*, 10.

Review

Biosynthesis of Metal Nanoparticles: Novel Efficient Heterogeneous Nanocatalysts

Jose M. Palomo [1,*] and Marco Filice [2]

1 Departament of Biocatalysis, Institute of Catalysis (CSIC), Marie Curie 2, Cantoblanco, Campus UAM, 28049 Madrid, Spain

2 Advanced Imaging Unit, Spanish National Research Center for Cardiovascular Disease (CNIC), 28049 Madrid, Spain; mfilice@icp.csic.es

* Correspondence: josempalomo@icp.csic.es; Tel.: +34-91-5854768; Fax: +34-91-5854670

Academic Editors: Hermenegildo García and Sergio Navalón
Received: 17 February 2016; Accepted: 26 April 2016; Published: 5 May 2016

Abstract: This review compiles the most recent advances described in literature on the preparation of noble metal nanoparticles induced by biological entities. The use of different free or substituted carbohydrates, peptides, proteins, microorganisms or plants have been successfully applied as a new green concept in the development of innovative strategies to prepare these nanoparticles as different nanostructures with different forms and sizes. As a second part of this review, the application of their synthetic ability as new heterogonous catalysts has been described in C–C bond-forming reactions (as Suzuki, Heck, cycloaddition or multicomponent), oxidations and dynamic kinetic resolutions.

Keywords: metal nanoparticle; biosynthesis; peptides; sugars; proteins; heterogeneous catalysis

---

1. Introduction

Nanotechnology has experimented a tremendous rise in the last decade [1–4]. In particular, the design of nanoparticles (NPs) has represented a very promising strategy alternative to conventional processes, especially of great application in environmental and biomedical problems (such as drug delivery, imaging, *etc.*) [5–8].

However, the field of nanocatalysis focused on the use of nanoparticles ashomogeneous or heterogeneous catalyst has been growth during the last years, [9–13]. The large surface-to-volume ratio of nanoparticles compared to bulk materials makes them attractive candidates for its use as catalysts. Especially the preparation and characterization of NPs from noble metals, which constitutes an important branch of heterogeneous catalysis in the chemical industry, represents an important challenge. The advantages of a very high superficial area make them excellent catalysts, reducing the amount of catalyst per gram of product making the process more sustainable. These NPs are synthesized by chemical methods in such a way to obtain good amounts, controlling the size and the form of the NPs [14–16]. However, in most cases hazardous conditions are used, toxic solvent, high amounts of energy (200 °C), which reduce a possible industrial implementation of this nanocatalyst. Most of these methods are still in the development stage and problems are often experienced with stability of the prepared nanoparticles, control of the crystal growth and aggregation of the particles [14–16]. Therefore the design of new synthetic approaches which considering an easy, rapid and sustainable strategy represents an important issue.

In recent years, a number of new green strategies have been described in literature. They are based on the capacity of biomolecules to induce the formation of these nanoparticles, sometimes even controlling the size and the structural form, and avoiding aggregation problems [17–27].

In this way, small molecules such as monosaccharides (glucose or galactose), or aminoacids and short peptides has been used as reducing agent for *in situ* creation of these metallic nanoparticles [17–21].

In addition, in a more precise way, the application of more complex biomolecules such as proteins or even microorganisms have been successfully applied to create nanoparticles and also hybrid systems with very high potential catalytic properties [23–26]. Bionanostructures, in which an enzyme is specifically encapsulated in a nanocluster or immobilized on biofunctionalized nanoparticles [10] are another category of catalysts with excellent features in cascade reactions. In particular, heterogeneous enzyme–metal nanoparticle nanohybrids with multiple catalytic activies are interesting in organic synthesis.

In this review, an overview of the recent advances on these new biosynthetic strategies and the use of the formed nanoparticles as catalyst in chemical processes is shown.

2. Synthesis of Metal Nanoparticles Induced by Glucosides

The most recent strategies targeting the synthesis of metallic nanoparticles have expected the introduction of green methodologies. In this sense, the use of glycosides avoids the necessity to apply toxic materials.

Thus, green synthesis of NPs induced by glucose and different glycopyranosides has been successfully reported [18,19,28]. In this case, very interesting results were found in the synthesis of gold nanoparticles (AuNPs) using eight different glucose derivatives [27].

A room temperature and easy synthetic method of AuNPs was developed using auric acid, sodium hydroxide and different glycosides as reducing agent (Figure 1).

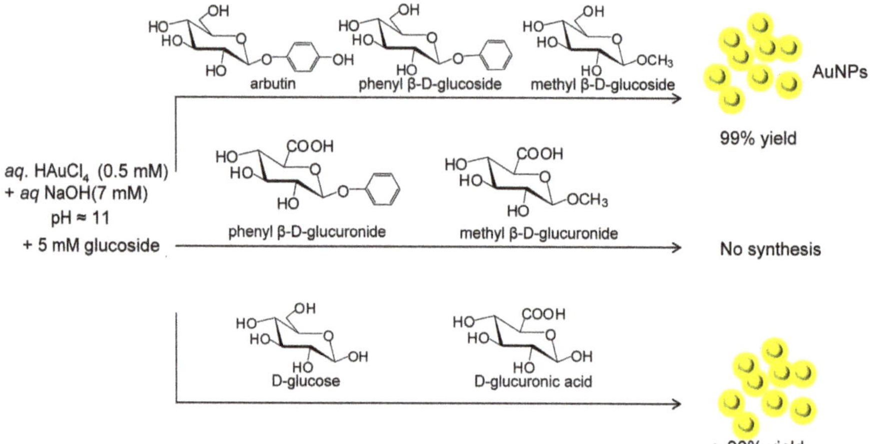

Figure 1. Synthesis of gold nanoparticles (AuNPs) induced by different glucose derivatives.

Eight sugar-containing reductants were used for comparison. *C-6* position of glycosides was oxidized to a carboxylic acid during the reduction of auric acid in the formation of AuNPs in the case of sugars substituted at the anomeric carbon/position (Figure 1). In the case of glucose or glucuronic acid (COOH in *C-6*), the NPs formation may be due to the oxidation of an aldehyde generated *via* anomerization. In this way, this explains why the synthesis in the presence of the glucuronic acid derivatives substituted at the anomeric position did not work (Figure 1).

Furthermore, significant differences in the final yield in AuNPs and especially the form and the size of the nanoparticles obtained by high resolution transmission electron microscope (HRTEM) (from 8 to 27 nm) depended on the sugar derivative (Figure 2).

The use of 1-phenyl or 1-methyl-β-glucopyranoside gave the highest synthetic yield (>99%) of homogeneous mono-dispersed round gold nanoparticles (13.15 nm, or 10.95 nm respectively), whereas using glucose or glucuronic acid the synthesized AuNPs showed multiple forms (16 and 8.8 nm). In the presence of arbutin, nanoparticles showed an amorphous form (27.4 nm) (Figure 2).

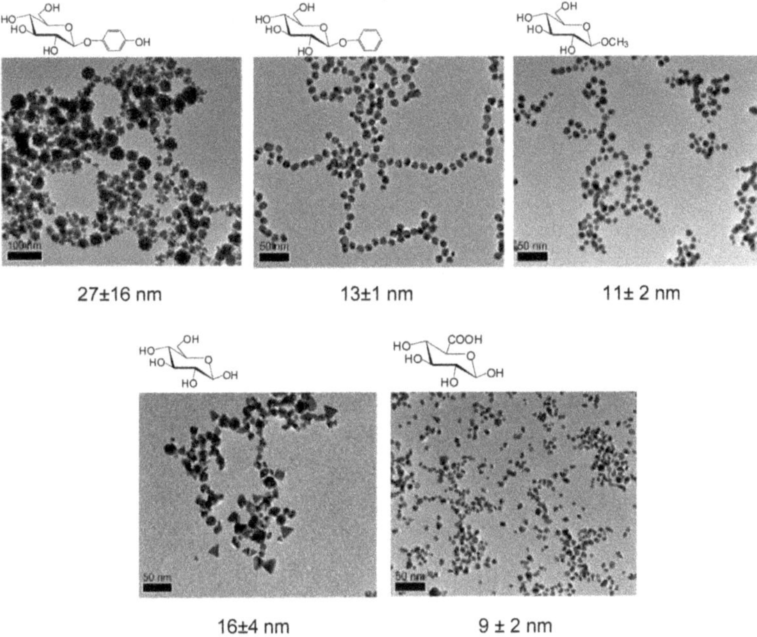

Figure 2. Characterization of the glucoside-induced AuNPs by high resolution transmission electron microscope (HRTEM).

Therefore, this work shows how the size and the form of gold nanoparticles can be controlled by using different sugars as additives. This aspect is of special interest for example in the case of ultrasmall nanoparticles, where their colloidal stability can be affected by these small size differences, underscoring the importance of particle uniformity in nanomedicine [28].

These strategies could be also extended with other metals and also using other sugars derivatives with different degree of hydrophobicity by substitution in the anomeric and other positions.

In a recent approach, AgNPs were synthesized using aqueous extract of *Clerodendron serratum* leaves, which have high contents of polyphenol glycosides and quercetin 3-*O*-β-D-glucoside [29].

This glucose derivative caused the reduction of the silver ions of silver nitrate in 30 min at room temperature, forming spherical AgNPs with the size range of 5–30 nm (Figure 3A). In addition, the strategy of using glycosides has been successfully used in the preparation of mesoporous nanostructures [30].

Mesoporous silica-coated silver nanoparticles (Ag@MSN) have been prepared by a two mild step synthesis. Glucose was used as reducing agent for silver ions whereas arginine was used for the formation of silica. The nanostructure presented single Ag nanoparticles as cores of diameter *ca.* 30 nm surrounded a silica shell (thickness of *ca.* 30–40 nm), with a total average particle size (*ca.* 110 nm) (Figure 3B).

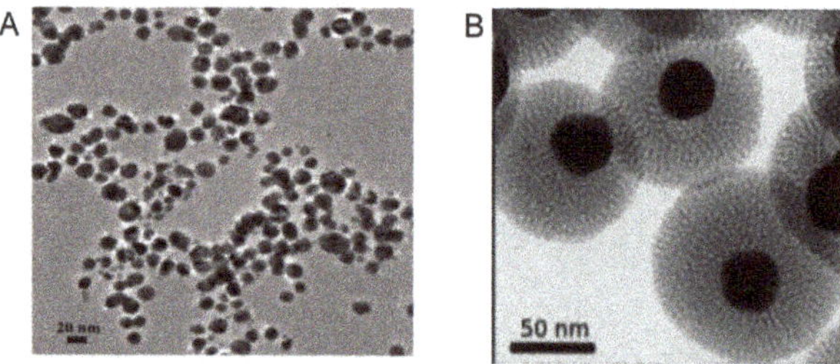

Figure 3. TEM image of different AgNPs. (**A**) AgNPs; (**B**) Mesoporous silica-coated silver nanoparticles (Ag@MSN) coated on a silicon substrate.

Another interesting example is the creation *in situ* of glycosylated functionalized gold nanoparticles [31].

In this case, the glycoside acted as reducing agent but also is a specific moiety for particle functionalization. A glycopolymer (cellulose) was activated in the anomeric position by thiosemicarbazide producing glucoside thiosemicarbazone (Figure 4A). This activated sugar combined with an aqueous *N*-methylmorpholine *N*-oxide (NMMO)-a molecule which permits the solubilization of the water-insoluble cellulose- and a dilute aqueous HAuCl solution finally producing glyco-AuNPs. TEM analysis confirmed the formation of AuNP aggregates with primary sizes of *ca.* 10–20 nm in diameter (Figure 4B). Same protocol was used to successfully synthesize glyco-AgNPs with distribution size of *ca.* 5–30 nm diameters.

The combination of NMMO-mediated GNP synthesis and immobilization of sugar reducing ends to an Au° matrix, allowed the design of a diverse array of carbohydrate-GNP conjugates by tailoring the functional sugars, e.g., cellulose, chitin, chitosan, maltose and lactose [18].

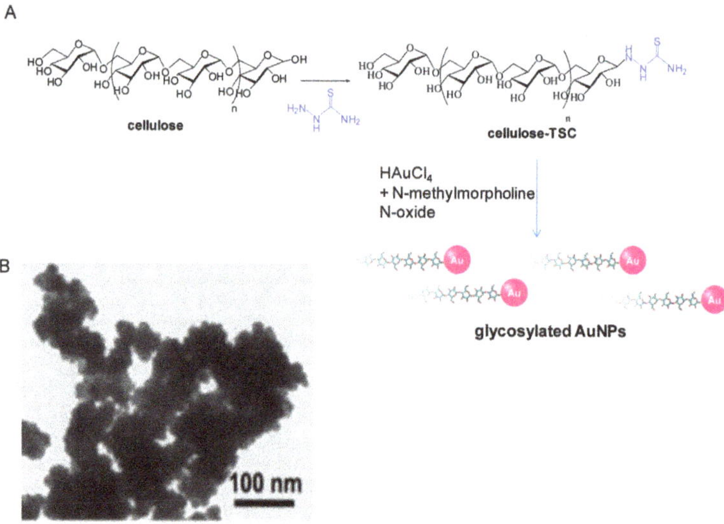

Figure 4. Synthesis of glycosylated AuNPs. (**A**) Synthetic scheme; (**B**) TEM image of glyco-AgNPs.

3. Biosynthesis of Metal Nanoparticles by Peptides

The synthesis of biocompatible metal nanoparticles can be performed by using peptides as multifunctional reagents (reducing and capping agents) under mild conditions [32–36]. The large diversity of peptides available provides a new opportunity to organize, interact, and direct the shape, size, and structure evolution of the metal nanoparticles in more varied and innovative ways. In some cases these peptides also show the capability to reduce metal ions and to template the crystal growth of the metal nanoparticles [37].

The use of a short conjugated peptide, such as a biotinylated di-tryptophan peptide was applied by Mishra and coworkers for the one-pot synthesis of stable gold nanoparticles [32]. The tryptophan dipeptide stabilized the NPs generated with an average size between 4 and 6 nm. Furthermore, self-assembled superior and ordered nanostructures of variable size were afforded where the AuNPs were scattered inside the biotinylated spherical scaffold in a controlled manner (Figure 5).

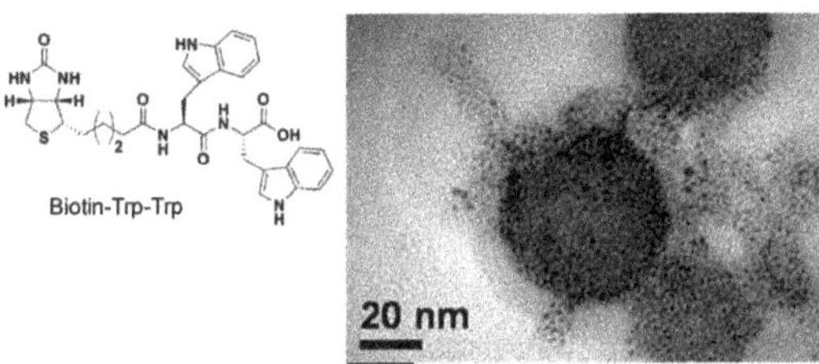

Figure 5. Self-organized AuNPs on the surfaces of biotin-Trp-Trp scaffold.

In another strategy, Giese *et al.* have recently described the synthesis of AgNPs under electron transfer conditions [33]. Ag^+ ions are bound by a peptide including a histidine (Figure 6A) as the Ag-binding amino acid, and a tyrosine as a photo inducible electron donor. The presence of chloride ions was necessary for the final formation of AgNPs, which occur on AgCl microcrystals in the peptide matrix. In this way, by controlling the irradiation times, the formation of Ag@AgCl/peptide nanocomposites with a sized of 100 nm at the beginning of the process was obtained, which are cleaved after time finally generating the AgNPs with a diameter of 15 nm (Figure 6).

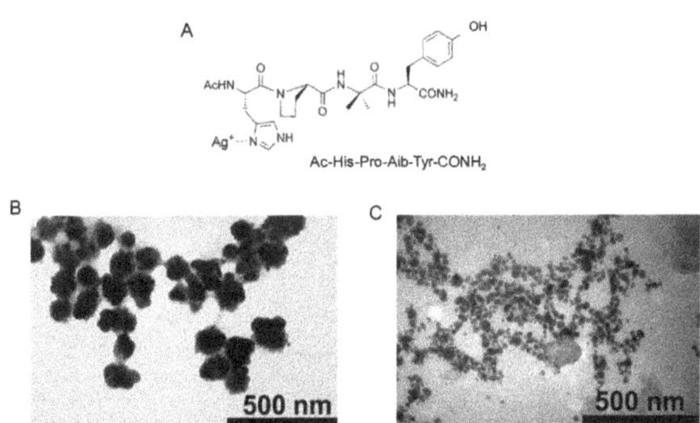

Figure 6. Synthesis of AgNPs. (**A**) Peptide structure; (**B**,**C**) TEM pictures Ag^+-peptide complex after different irradiation times, t = 30 s and t = 30 min, respectively.

Tekinay and coworkers described the design and application of a multidomain peptide for single-step, size-controlled synthesis of biofunctionalized AuNPs (Figure 7) [34]. Size-controlled synthesis of AuNPs with this peptide was possible due to the 3,4-dihydroxy-L-phenylalanine (L-DOPA) functional group, a residue known for its reductive role. The authors showed DOPA coupled its oxidation to the reduction of Au (III) ions, thereby leading to the formation of biofunctionalized AuNPs. Hence, the DOPA-mediated peptide design enables concerted one-pot reduction, stabilization and functionalization of resulting AuNPs whereby no additional reagent or reaction is needed.

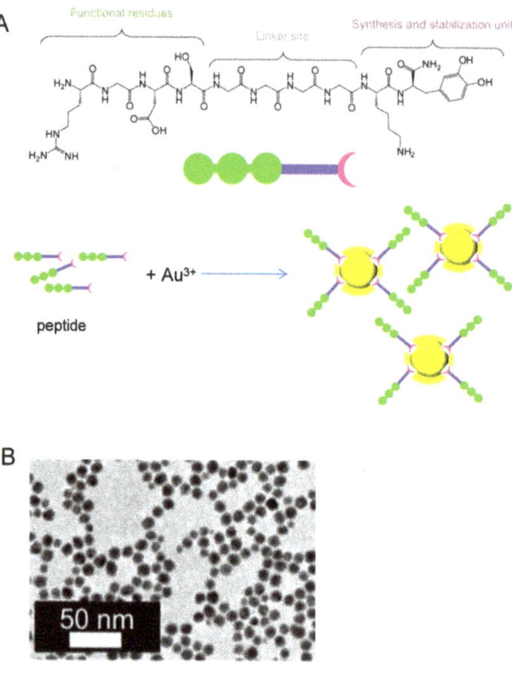

Figure 7. Synthesis of Biofunctionalized AuNPs. (**A**) Peptide and scheme of NPs formation; (**B**) TEM images of the AuNPs.

4. Bio-Inspired Synthesis of Nanoparticles by Proteins

Among these strategies, the employment of microbial enzymes for nanoparticle synthesis is a new field with growing importance. In fact, considering that various enzymes have different capacities for synthesis of nanoparticles in a wide set of shapes and sizes, it is very important to find suitable enzymes for such purposes and improve the method for optimal nanoparticle production. Conversely, nanoparticles obtained by cell-based methods forcefully bind to the microbial biomass resulting in high-cost laborious steps of separation and purification of nanoparticles from microbial cells. In this line, many examples have been reported in literature. For example, Cholami-Shabami *et al.* developed a cell-free viable approach for synthesis of gold nanoparticles using NADPH-dependent sulfite reductase purified from *Escherichia coli.* [38] Highly stable gold nanoparticles were produced by reductive process after application of the sulfite reductase to an aqueous solution of $AuCl_4^-$. The enzymatically synthesized gold nanoparticles showed strong inhibitory effect towards the growth of various human pathogenic fungi [38]. Another interesting approach was developed by Kas and coworkers permitting the achievement of nanosilica-supported Ag nanoparticles by means of a biosynthetic protocol (Figure 8) [39]. After immobilizing a protein extract proceeding from *Rhizopus oryzae* on the surface of a nanosilica structured support, the authors carried out the synthesis of AgNPs using this biohybrid as host for the growth of AgNPs on its surface. In the proposed *in situ* synthetic process, the Ag^+ ions-proceeding from an $AgNO_3$ solution as metal precursor-were considered to be rapidly adsorbed on negatively charged protein surfaces through electrostatic interaction. The reduction of Ag^+ ions and the subsequent formation of AgNPs was due to the electron transfer between the metal ions and the functional groups of proteins.

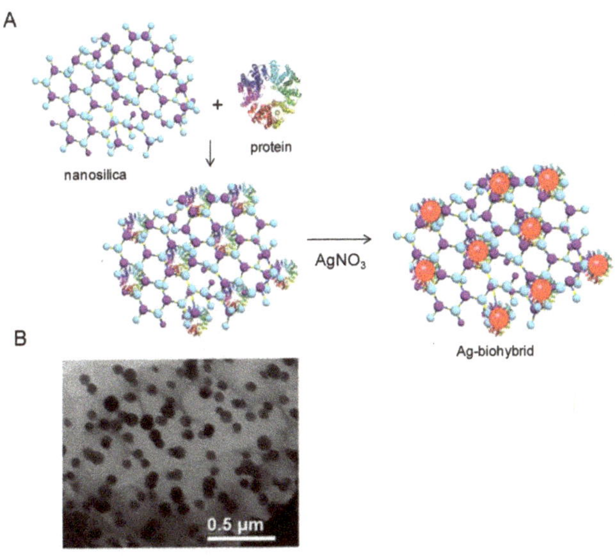

Figure 8. Synthesis of Ag-biohybrid. (**A**) Scheme of the formation of the nanostructure; (**B**) TEM of Ag-nanohybrid. Reproduced with permission from [39]. Copyright the Royal Society of Chemistry, 2013.

Following this research line, a novel type of heterogeneous hybrid nanocatalysts, composed by metal NPs embedded in an enzymatic net, was generated *in situ* under very mild reaction conditions from the simple mixture of lipase from *Candida antarctica* fraction B (CAL-B) with a homogeneous aqueous solution of a noble metal salt (Ag^{+}, Pd^{2+}, or Au^{3+}) (Figure 9) [22,40].

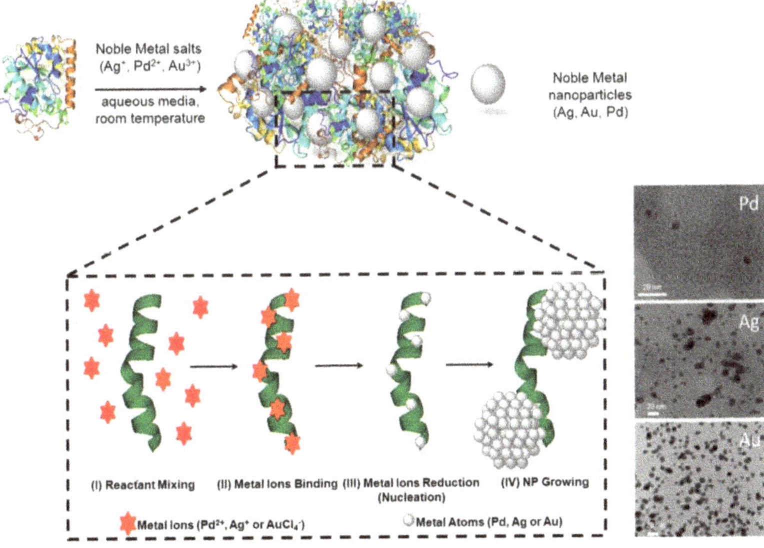

Figure 9. Preparation of metal bionanohybrids.

This new hybrid nanocatalysts combines metallic and enzymatic catalytic activitie. The use of an enzyme in the methodology permitted the generation of small metal NPs (e.g., around 2 nm core size for Pd NPs) without the need for any external reducing agent, exploiting the reductive ability of the biomacromolecule (biomineralization), which moreover remains catalytically active at the end of the synthesis [22,40]. Based on the same general bio-based strategy, Das *et al.* reported an interesting biosynthetic route for cost-effective productions of various metal NPs (Pd, Pt, and Ag) on the surface of fungal mycelia [41]. The metal NPs were synthesized through an electrostatic interaction of metal ion precursors, followed by their reduction to nanoparticles by surface proteins finally decorating the mycelia surface in a homogeneous way. It results worth of note as, by means of this strategy, the size and shape varied depending on the metal NPs. In fact, "flower"-like branched nanoparticles were obtained in the case of Pd and Pt, while Ag produced spheroidal nanoparticles, this structural characteristic is a key-element of their catalytic activity which is assessed in hydrogenation and Suzuki C–C coupling reactions in aqueous solution [41].

Even engineered proteins were revealed to be useful in the synthesis of precise and highly functionalized metal nanoparticles. For example, a small variant of protein A has been used as biotemplate in the one-step synthesis and biofunctionalization of AuNPs [42]. This biotemplate is composed by a thiolate ligand capable of interacting with the AuNP surface and controlling the nanoparticle nucleation and growth, thus allowing the nanoparticle size to be finely tuned. This crucial feature clearly resulted as key-advantage of this approach, which allow for high-quality AuNPs to be obtained in the water phase, and therefore avoiding the transfer from organic solvents, which usually results in a lack of long-term stability [42].

Moving a step over the well-known strategy based on the creation of metallic nanoparticles inside the cavity of hollow protein (*i.e.*, ferritin), Jang and coworkers described the synthesis of thin-walled (*ca.* 40 nm) SnO_2 nanotubes functionalized with catalytic Pt and Au nanoparticles via a protein templating route [43]. After the creation of metal NPs inside the cavity of an apoferritin template via $NaBH_4$ reductive strategy starting from metal salt precursors, as the prepared hybrids catalysts were used to decorate both the interior and exterior surfaces of the thin-walled SnO_2 nanotubes. After calcination, the protein cage was eliminated leaving a well-dispersed layer of catalytic metal nanoparticles immobilized on nanotubes surface. Such a uniform surface distribution, resulting from the repulsion between the proteic cages before their calcination, granted a final very high surface area-to-volume ratio leading to superior catalytic performances for example in gas sensing [43].

5. Biosynthesis of Metal Nanoparticles by Microorganisms

Apart of the use of small molecules or even proteins, the use of entire biological units as prokaryotic or eukaryotic microorganisms have been employed for the preparation of nanoparticles of different metals (Au, Ag, Cd, Pt, Zn, Fe_3O_4) under moderate pressures and temperatures [44–47].

Microorganisms are capable of adsorbing and accumulating metals. They also secrete large amounts of enzymes, which are involved in the enzymatic reduction of metals ions [48,49]. Microbial synthesis of metallic nanoparticles can take place either outside or inside the cell [44], producing metal NPs, which have characteristic features similar to nanomaterials, which are synthesized chemically [14]. The localization, size or shape of the nanoparticles depend on the microorganism specie used [44].

In this way, the production of metal nanoparticles by fungi is one of the most successful strategies [42,50–53].

For example, in one of the cases, the fungus *Aspergillus japonica* was used for the reduction of Au (III) into Au NPs. Spherical and well distributed on fungal mycelia particles were found. The size of the particles ranges predominantly between 15 and 20 nm. Furthermore, the nanoparticles were simultaneously immobilized on the fungus surface creating a heterogeneous hybrid with interesting catalytic properties [50].

Another example of the use of fungus was described by Loshchinina and coworkers [51]. The authors described the synthesis of AuNPs by the fungus *Basidiomycete lentinus edodes*. TEM experiment demonstrated the formation of spherical Au(0) nanoparticles inside the mycelia cells mostly of 5 to 15 nm with minor part of 30 to 50 nm diameter. An Au distribution map was obtained that supported these electron-dense formations to be intracellular Au nanoparticles.

This is also the first time that fungal intracellular phenol-oxidizing enzymes (laccases, tyrosinases, and Mn-peroxidases) have been involved in Au reduction to give electrostatically stabilized colloidal solutions.

Also the preparation of AuNPs has been described by Gupta and coworkers using in this case the fungus *Trichoderma* sp. [52]. The biosynthesis of the nanoparticles was rapid at 30 °C using cell-free extract of the *Trichoderma viride*, producing extracellular AuNPs with particle size of 20–30 nm. Using *Hypocrea lixii* the synthesis was similar but at 100 °C, obtaining smaller nanoparticles (<20 nm).

The use of recombinant *E. coli* expressing a tyrosinase from *Rhizobium etli* has been described as interesting green strategy to synthesize gold nanoparticles [54]. Tyrosinase is an important enzyme in biology involved in production of melanin. The catalytic function is the oxidation of L-tyrosine to 3-(3,4-dihydroxyphenyl)-L-alanine (L-DOPA) and further to dopaquinone and melanin. In particular, eumelanin–natural pigment, which contains carboxyl, amine, hydroxyl groups, quinone and semiquinone groups–was used as agent to reduce the metals ions. In the presence of L-DOPA and gold ions, exogenous AuNPs were formed (Figure 10A). The absence of L-DOPA failed in the formation of AuNPs, demonstrating that the presence of eumelanin (generate by the enzyme with L-DOPA) is critical for the nanoparticles formation (Figure 10B). In the absence of gold ions, the transformation of L-DOPA to eumelanin was observed (Figure 10C). The TEM analysis demonstrated that the AuNPs showed a particle size average of around 12 nm (Figure 10D). The strategy was successfully applied to other metals obtaining NPs with a particle size between 7 and 13 nm [53].

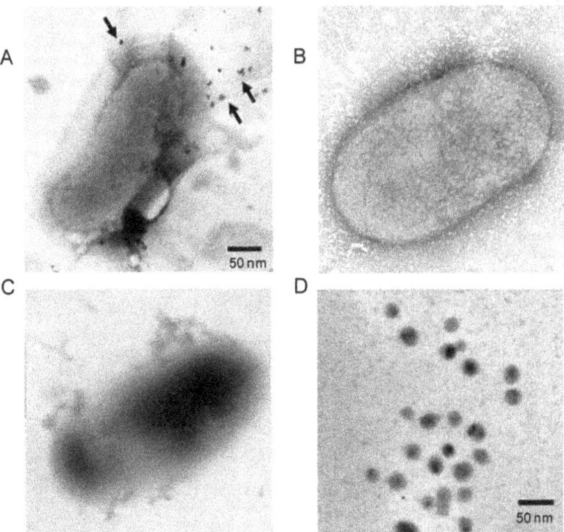

Figure 10. Eumelanine from tyrosinase induced the synthesis of gold nanoparticles. (**A**) TEM image of *R. etli* cell in the presence of 3-(3,4-dihydroxyphenyl)-L-alanine (L-DOPA) and Au ions; (**B**) TEM image of *R. etli* cell in the presence of Au ions; (**C**) TEM image of *R. etli* cell in the presence of L-DOPA; (**D**) TEM of the synthesized AuNPs. Reproduced with permission from [53]. Copyright the Royal Society of Chemistry, 2014

6. Biosynthesis of Metal Nanoparticles by Plant Extracts

Another strategy used for preparation of biosynthetic green metal nanoparticles is based on the use of extract of different plants. These contain a wide amount of natural products such as polyphenylols, alkaloids, flavonoids and terpenoids (reducing and stabilizing agents), which induced the final formation of the nanoparticles. An interesting review article about that has been published recently by Banerjee and coworkers [24] and no extended description of this strategy will be present here.

However, we would like to emphasize two very recent works in the preparation of mono and bimetallic nanoparticles [12,54].

The first describes the synthesis of platinum nanoparticle (PtNP) aqueous colloid by utilizing black wattle tannin (BWT), a typical plant polyphenol (Figure 11) [54]. The hydroxyl groups of this molecule act as reducing agent but also as stabilizers and protecting the PtNPs from deactivation caused by oxygen atmosphere. The aromatic framework of BWT prevents the aggregation of the nanoparticles in contrast with the use of the most hydrophilic organic small molecules. Stable heterogeneous metallic nanoparticles under mild conditions can be then obtained by this method. Different amount of BWT were tested showing that only when 15 mg of BWT was added, completely monodispersed Pt small nanoparticles (*ca.* 1.8 nm diameter size) were formed (Figure 11B).

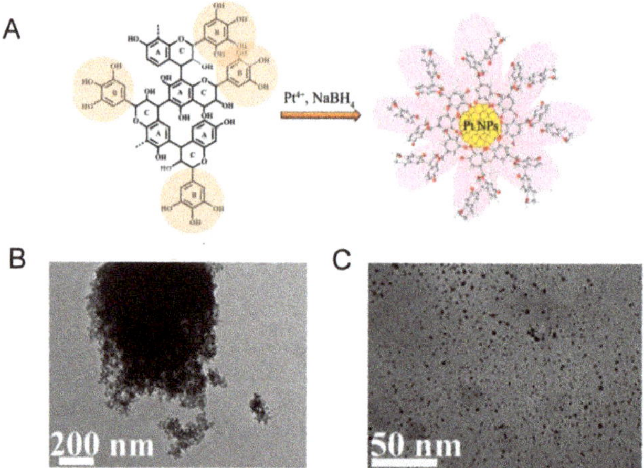

Figure 11. Synthesis of Pt nanoparticles induced by black wattle tannin (BWT). (**A**) Scheme of the formation of PtNPs by BWT; (**B**) TEM image of synthesized PtNPs using 2 mg of BWT; (**C**) TEM image of synthesized PtNPs using 15 mg of BWT. Reproduced with permission from [54]. Copyright the Royal Society of Chemistry, 2016.

The second example refers to the synthesis of Fe, Pd and Fe-Pd bimetallic nanoparticles using the medicinally potent aqueous bark extract of *Ulmus davidiana* [12]. As shown in the previous case, the polyols present in the plant were responsible for reducing and capping of the nanoparticles. The NPs preparation was achieved using aqueous solution of *Ulmus* adding firstly Fe_2O_3, $PdCl_2$ or the bimetallic iron oxide and then the palladium chloride at 60 °C. The three sort NPs showed different diameter size determined by TEM. Spherical FeNPs showed 50 nm or PdNPs with 5 nm-sized were obtaining using a unique metal in the synthesis. On the bimetallic, the PdNPs were adsorbed on the outer surface of the FeNPs (Figure 12A). The distribution size were 3–7 nm of Pd clustered around 30–70 nm of FeNPs (Figure 12B).

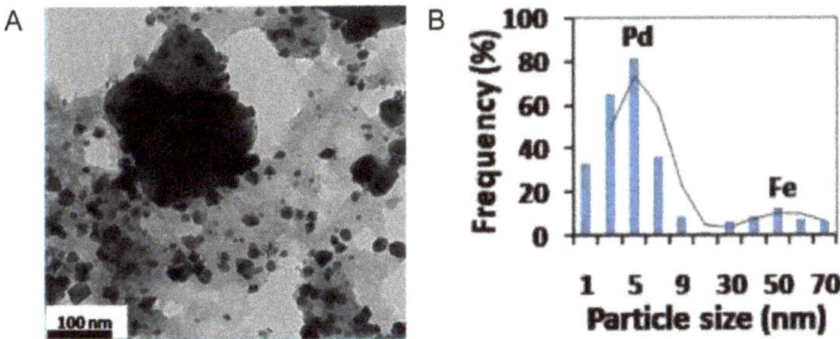

Figure 12. Characterization of Fe-Pd NPs synthesized by *Ulmus davidiana*. (**A**) TEM image of Fe-Pd NPs; (**B**) Particle size distribution histogram. Reproduced with permission from [12]. Copyright the Royal Society of Chemistry, 2015.

7. Application of the Biosynthesized Metallic NPs as Nanocatalysts

Beside the description of the metal NPs preparation, the different methods presented above have been tested on the reduction of *p*-nitrophenol to *p*-aminophenol as general model reaction in order to assess the catalytic properties of generated nanobiocatalysts.

As representative example, the catalytic constant (k) and the turnover frequency (TOF) of that reaction using the CAL-B-Pd biohybrid were calculated, retrieving excellent k and TOF values (0.6 min^{-1} and almost 150 min^{-1}, respectively). The TOF value was the highest described in the literature for this reaction at that moment [22].

In parallel to the model aryl-amine synthesis, some of the previous synthesized heterogeneous catalysts have been successfully used in different typology of complex reactions.

Thus, the bimetallic nanoparticles formed by Fe and Pd represented a catalyst magnetically recyclable and reused in [3 + 2] cycloaddition reaction [12] (Figure 13). These Fe-Pd NPs displayed better catalytic activity, with final isolated yields from 89% to 98% for the synthesis of 12 different naphtha[1,2-b]furan-3-carboxamides and benzofuran-3-carboxamides compared to their respective monometallic nanoparticles (Figure 13).

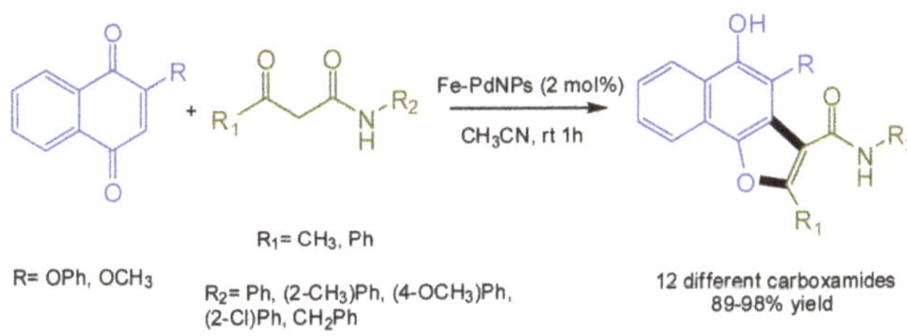

Fe-PdNPs (2 mol%)
CH_3CN, rt 1h
R= OPh, OCH_3
R_1= CH_3, Ph
R_2= Ph, (2-CH_3)Ph, (4-OCH_3)Ph, (2-Cl)Ph, CH_2Ph
12 different carboxamides
89-98% yield

Figure 13. Synthesis of substituted carboxamides catalyzed by Fe-Pd NPs nanocatalyst.

Advantages of this heterogeneous nanocatalyst were an easy recovery using an external magnetic field, and recycling (maintaining almost complete activity after 5 times).

Another interesting example of practical usefulness is represented by the BWT-Pt colloid catalyst [54]. This nanocatalyst showed interesting activity in a series of biphasic oxidation of aromatic alcohols and aliphatic alcohols under mild aerobic conditions in aqueous media. The results were from moderate yields, for alkyl compounds, to high yield, for aromatic alcohols. The best results were obtained in the oxidation of phenylmethanol and 1-phenylethanol with more than 80% yield of product (Figure 14). The nanocatalyst was very stable and no decrease in the oxidative activity was observed after seven cycles.

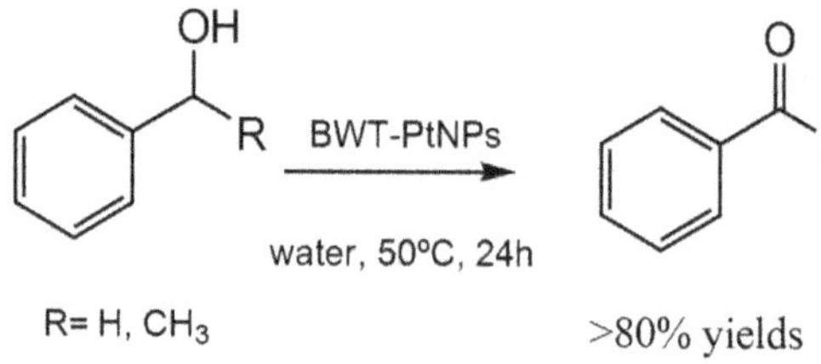

Figure 14. Oxidation of phenylalcohols by BWT-PtNPs.

The synthesis of propargylamines by a multi-component reaction has been successfully catalyzed by means of the gold-NPs-fungal hybrid bionanocatalyst [46]. These bioNPs produced different propargylamines by an A3 coupling process, a very interesting synthetic way to produce heterocyclic frameworks [55].

The high versatility of the nanocatalyst was demonstrated and over 80% yields of various propargylamine derivatives were obtained using a variety of aromatic aldehydes (R = H, CH_3, Cl) coupled with secondary amine and alkynes after 24 h (Figure 15). In addition, this heterogeneous hybrid showed good recyclability.

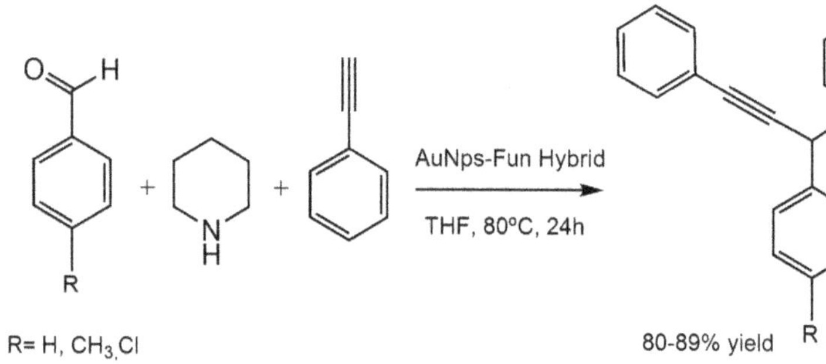

Figure 15. Multicomponent reaction catalyzed by AuNPs Hybrid.

The Pd-lipase bionanohybrid described before was successfully applied as excellent heterogeneous catalysts in C–C bond reaction [22]. This resulted in an extremely active Pd-catalyst in the Suzuki reaction, forming the biphenyl in >99% conversion using *ppb of catalyst* (Figure 16). Furthermore, the reaction was performed in pure water so making this organic reaction a green chemical process. A recyclable catalyst was obtained, used five times conserving almost the activity intact.

The particular advantage of this catalyst is that it conserved the native enzymatic activity, so it was successfully applied in the dynamic kinetic resolution of *rac*-phenylethylamine in organic solvent, a tandem catalytic process (both enzymatic and Pd catalysis acting at the same time) (Figure 16). The bionanohybrid quantitatively produced the enantiopure (*R*)-benzylamide with *ee* > 99%. Even in this case, the recyclability of the catalyst was excellent.

Finally, also the previously described "flower"-like branched Pd nanoparticles were used as heterogeneous catalyst in the Suzuki reaction of formation of biaryl with satisfactory results (99% conversion) [41].

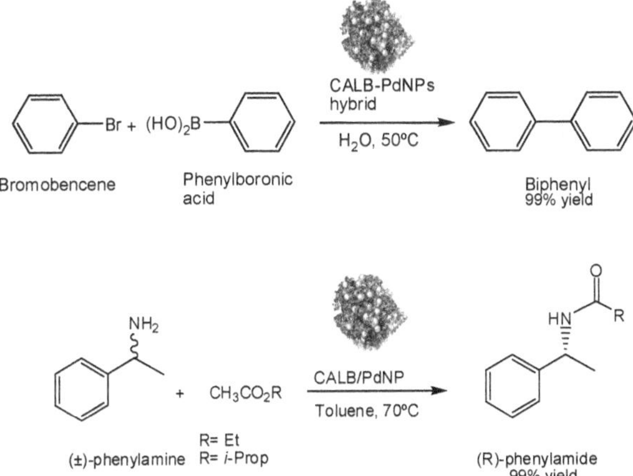

Figure 16. Catalytic applications of CAL-B-PdNPs hybrid.

8. Conclusions

In conclusion, this review showed the most recent different strategies used for the biosynthesis of metal nanoparticles. The application of biological entities represents an interesting and green solution for the environmentally friendly synthesis of these nanoparticles. The use of enzymes as biomolecule permits the design of more precise nanostructures for many interesting chemical applications and it allows combination of one or more metallic activities together with enzymatic catalytic ones (excellent for cascade processes). Therefore, the development of newer and more efficient bio-methodologies for the creation of nanobiohybrids will be a future issue. Together with that, the reviewed examples showed also the excellent catalytic application of these heterogeneous nanocatalysts demonstrating the tremendous potential of their use in organic synthesis.

Acknowledgments: This work was supported by the Spanish National Research Council (CSIC). The author thanks the Ramon Areces Foundation for financial support. M.F also thanks the MINECO for the research grant SAF2014-59118-JIN (Programa Estatal de Investigación, Desarrollo e Innovación Orientada a los Retos de la Sociedad 2014: "Proyectos de I+D+i para jóvenes investigadores"), and co-funding by Fondo Europeo de Desarrollo Regional (FEDER).

Conflicts of Interest: The authors declare no conflict of interest.

References

1. Zhang, F.; Nangreave, J.; Liu, Y.; Yan, H. Structural DNA nanotechnology: State of the art and future perspective. *J. Am. Chem. Soc.* **2014**, *136*, 11198–11211. [CrossRef] [PubMed]
2. Mo, R.; Jiang, T.; Di, J.; Tai, W.; Gu, Z. Emerging micro- and nanotechnology based synthetic approaches for insulin delivery. *Chem. Soc. Rev.* **2014**, *43*, 3595–3629. [CrossRef] [PubMed]
3. Han, X.; Zheng, Y.; Munro, C.J.; Ji, Y.; Braunschweig, A.B. Carbohydrate nanotechnology: Hierarchical assembly using nature's other information carrying biopolymers. *Curr. Opin. Biotechnol.* **2015**, *34*, 41–47. [CrossRef] [PubMed]
4. Dai, Y.; Wang, Y.; Liu, B.; Yang, Y. Metallic nanocatalysis: An accelerating seamless integration with nanotechnology. *Small* **2015**, *11*, 268–289. [CrossRef] [PubMed]
5. Zhou, W.; Gao, X.; Liu, D.; Chen, X. Gold Nanoparticles for *in Vitro* Diagnostics. *Chem. Rev.* **2015**, *115*, 10575–10636. [CrossRef] [PubMed]
6. DaCosta, M.V.; Doughan, S.; Han, Y.; Krull, U.J. Lanthanide upconversion nanoparticles and applications in bioassays and bioimaging: A review. *Anal. Chim. Act.* **2014**, *832*, 1–33. [CrossRef] [PubMed]

7. Shi, D.; Sadat, M.E.; Dunn, A.W.; Mast, D.B. Photo-fluorescent and magnetic properties of iron oxide nanoparticles for biomedical applications. *Nanoscale* **2015**, *7*, 8209–8232. [CrossRef] [PubMed]
8. Goldberg, M.S. Immunoengineering: How nanotechnology can enhance cancer immunotherapy. *Cell* **2015**, *161*, 201–204. [CrossRef] [PubMed]
9. Dong, X.-Y.; Gao, Z.-W.; Yang, K.-F.; Zhang, W.-Q.; Xu, L.-W. Nanosilver as a new generation of silver catalysts in organic transformations for efficient synthesis of fine chemicals. *Catal. Sci. Technol.* **2015**, *5*, 2554–2574. [CrossRef]
10. Filice, M.; Palomo, J.M. Cascade reactions catalyzed by bionanostructures. *ACS Catal.* **2014**, *4*, 1588–1598. [CrossRef]
11. Serna, P.; Corma, A. Transforming nano metal nonselective particulates into chemoselective catalysts for hydrogenation of substituted nitrobenzenes. *ACS Catal.* **2015**, *5*, 7114–7121. [CrossRef]
12. Mishra, K.; Basavegowda, N.; Lee, Y.R. Biosynthesis of Fe, Pd, and Fe–Pd bimetallic nanoparticles and their application as recyclable catalysts for [3 + 2] cycloaddition reaction: A comparative approach. *Catal. Sci. Technol.* **2015**, *5*, 2612–2621. [CrossRef]
13. Aditya, T.; Pal, A.; Pal, T. Nitroarene reduction: A trusted model reaction to test nanoparticle catalysts. *Chem. Commun.* **2015**, *51*, 9410–9431. [CrossRef] [PubMed]
14. MubarakAli, D.; Gopinath, V.; Rameshbabu, N.; Thajuddin, N. Synthesis and characterization of CdS nanoparticles using C-phycoerythrin from the marine cyanobacteria. *Mater. Lett.* **2012**, *74*, 8–11. [CrossRef]
15. Saldan, I.; Semenyuk, Y.; Marchuk, I.; Reshetnyak, O. Chemical synthesis and application of palladium nanoparticles. *J. Mat. Sci.* **2015**, *50*, 2337–2354. [CrossRef]
16. Gutiérrez, L.; Costo, R.; Grüttner, C.; Westphal, F.; Gehrke, N.; Heinke, D.; Fornara, A.; Pankhurst, Q.A.; Johansson, C.; Veintemillas-Verdaguer, S.; *et al.* Synthesis methods to prepare single- and multi-core iron oxide nanoparticles for biomedical applications. *Dalton Trans.* **2015**, *44*, 2943–2952. [CrossRef] [PubMed]
17. Shervani, Z.; Yamamoto, Y. Carbohydrate-directed synthesis of silver and gold nanoparticles: Effect of the structure of carbohydrates and reducing agents on the size and morphology of the composites. *Carbohydr. Res.* **2011**, *346*, 651–658. [CrossRef] [PubMed]
18. Yokota, S.; Kitaoka, T.; Opietnik, M.; Rosenau, T.; Wariishi, H. Synthesis of gold nanoparticles for *in situ* conjugation with structural carbohydrates. *Angew. Chem. Int. Ed.* **2008**, *47*, 9866–9869. [CrossRef] [PubMed]
19. Engelbrekt, C.; Sørensen, K.H.; Zhang, J.; Welinder, A.C.; Jensen, P.S.; Ulstrup, J. Green synthesis of gold nanoparticles with starch–glucose and application in bioelectrochemistry. *J. Mater. Chem.* **2009**, *19*, 7839–7847. [CrossRef]
20. Care, A.; Bergquist, P.L.; Sunna, A. Solid-binding peptides: Smart tools for nanobiotechnology. *Trends Biotechnol.* **2015**, *33*, 259–268. [CrossRef] [PubMed]
21. Tan, Y.N.; Lee, J.Y.; Wang, D.I.C. Uncovering the design rules for peptide synthesis of metal nanoparticles. *J. Am. Chem. Soc.* **2010**, *132*, 5677–5686. [CrossRef] [PubMed]
22. Filice, M.; Marciello, M.; Morales, M.P.; Palomo, J.M. Synthesis of heterogeneous enzyme-metal nanoparticle biohybrids in aqueous media and their applications in C–C bond formation and tandem catalysis. *Chem. Commun.* **2013**, *49*, 6876–6878. [CrossRef] [PubMed]
23. Mittal, A.K.; Chisti, Y.; Banerjee, U.C. Synthesis of metallic nanoparticles using plant extracts. *Biotechnol. Adv.* **2013**, *31*, 346–356. [CrossRef] [PubMed]
24. Chinnadayyala, S.R.; Santhosh, M.; Singh, N.K.; Goswami, P. Alcohol oxidase protein mediated *in situ* synthesized and stabilized gold nanoparticles for developing amperometric alcohol biosensor. *Biosen. Bioelec.* **2015**, *69*, 151–161. [CrossRef] [PubMed]
25. Hulkoti, N.I.; Taranath, T.C. Biosynthesis of nanoparticles using microbes—A review. *Colloids Surf. B* **2014**, *121*, 474–483. [CrossRef] [PubMed]
26. Mashwani, Z.-U.-R.; Khan, T.; Khan, M.A.; Nadhman, A. Synthesis in plants and plant extracts of silver nanoparticles with potent antimicrobial properties: Current status and future prospects. *App. Microb. Biotechnol.* **2015**, *99*, 9923–9934. [CrossRef] [PubMed]
27. Sousa, A.A.; Hassan, S.A.; Knittel, L.L.; Balbo, A.; Aronova, M.A.; Brown, P.H.; Schuck, P.; Leapman, R.D. Biointeractions of ultrasmall glutathione-coated gold nanoparticles: Effect of small size variations. *Nanoscale* **2016**, *8*, 6577–6588. [CrossRef] [PubMed]

28. Jung, J.; Park, S.; Hong, S.; Ha, M.W.; Park, H.-G.; Lee, H.-J.; Park, Y.; Park, Y. Synthesis of gold nanoparticles with glycosides: Synthetic trends based on the structures of glycones and aglycones. *Carbohydr. Res.* **2014**, *386*, 57–61. [CrossRef] [PubMed]
29. Raman, R.P.; Parthiban, S.; Srinithya, B.; Vinod, V.; Savarimuthu, K.; Anthony, P.; Sivasubramanian, A.; Muthuraman, M.S. Biogenic silver nanoparticles synthesis using the extract of the medicinal plant *Clerodendron serratum* and its *in vitro* antiproliferative Activity. *Mat. Lett.* **2015**, *160*, 400–403. [CrossRef]
30. Saint-Cricq, P.; Wang, J.; Sugawara-Narutaki, A.; Shimojima, A.; Okubo, T. A new synthesis of well-dispersed, core–shell Ag@SiO mesoporous nanoparticles using amino acids and sugars. *J. Mater. Chem. B* **2013**, *1*, 2451–2454. [CrossRef]
31. Kitaoka, T.; Yokota, S.; Opietnik, M.; Rosenau, T. Synthesis and bio-applications of carbohydrate–gold nanoconjugates with nanoparticle and nanolayer forms. *Mat. Sci. Eng. C* **2011**, *31*, 1221–1229. [CrossRef]
32. Mishra, N.K.; Kumar, V.; Joshi, K.B. Fabrication of gold nanoparticles on biotin-ditryptophan scaffold for plausible biomedical applications. *RSC Adv.* **2015**, *5*, 64387–64394. [CrossRef]
33. Kracht, S.; Messerer, M.; Lang, M.; Eckhardt, S.; Lauz, M.; Grobty, B.; Fromm, K.M.; Giese, B. Electron Transfer in Peptides: On the Formation of Silver Nanoparticles. *Angew. Chem. Int. Ed.* **2015**, *54*, 2912–2916. [CrossRef] [PubMed]
34. Gulsuner, H.U.; Ceylan, H.; Guler, M.O.; Tekinay, A.B. Multi-domain short peptide molecules for *in situ* synthesis and biofunctionalization of gold nanoparticles for integrin-targeted cell uptake. *ACS Appl. Mater. Interfaces* **2015**, *7*, 10677–10683. [CrossRef] [PubMed]
35. Belser, K.; Slenters, T.V.; Ofumbidzai, C.; Upert, G.; Mirolo, L.; Fromm, K.M.; Wennermers, H. Silver nanoparticle formation in different sizes induced by peptides identified within split-and-mix libraries. *Angew. Chem. Int. Ed.* **2009**, *48*, 3661–3664. [CrossRef] [PubMed]
36. Tomizaki, K.-Y.; Kubo, S.; Ahn, S.-A.; Satake, M.; Imai, T. Biomimetic alignment of zinc oxide nanoparticles along a peptide nanofiber. *Langmuir* **2012**, *28*, 13459–13466. [CrossRef] [PubMed]
37. Naik, R.R.; Stringer, S.J.; Agarwal, G.; Jones, S.E.; Stone, M.O. Biomimetic synthesis and patterning of silver nanoparticles. *Nat. Mater.* **2002**, *1*, 169–172. [CrossRef] [PubMed]
38. Gholami-Shabani, M.; Shams-Ghahfarokhi, M.; Gholami-Shabani, Z.; Akbarzadeh, A.; Riazi, G.; Ajdari, S.; Amani, A.; Razzaghi-Abyaneh, M. Enzymatic synthesis of gold nanoparticles using sulfite reductase purified from Escherichia coli: A green eco-friendly approach. *Process Biochem.* **2015**, *50*, 1076–1085. [CrossRef]
39. Das, S.K.; Khan, M.R.; Guhab, A.K.; Naskar, N. Bio-inspired fabrication of silver nanoparticles on nanostructured silica: Characterization and application as a highly efficient hydrogenation catalyst. *Green Chem.* **2013**, *15*, 2548–2557. [CrossRef]
40. Cuenca, T.; Filice, M.; Palomo, J.M. Palladium nanoparticles enzyme aggregate (PANEA) as efficient catalyst for Suzuki-Miyaura reaction in aqueous media. *Enzyme Microb. Technol.* **2016**, in press. [CrossRef]
41. Das, S.K.; Parandhaman, T.; Pentela, N.; Islam, A.K.M.M.; Mandal, A.B.; Mukherjee, M. Understanding the biosynthesis and catalytic activity of Pd, Pt, and Ag nanoparticles in hydrogenation and Suzuki coupling reactions at the nano–bio interface. *J. Phys. Chem. C* **2014**, *118*, 24623–24632. [CrossRef]
42. Colombo, M.; Mazzucchelli, S.; Collico, V.; Avvakumova, S.; Pandolfi, L.; Corsi, F.; Porta, F.; Prosperi, D. Protein-assisted one-pot synthesis and biofunctionalization of spherical gold nanoparticles for selective targeting of cancer cells. *Angew. Chem. Int. Ed.* **2012**, *51*, 9272–9275. [CrossRef] [PubMed]
43. Jang, J.-S.; Kim, S.-J.; Choi, S.-J.; Kim, N.-H.; Hakim, M.; Rothschild, A.; Kim, I.-D. Thin-walled SnO_2 nanotubes functionalized with Pt and Au catalysts via the protein templating route and their selective detection of acetone and hydrogen sulfide molecules. *Nanoscale* **2015**, *7*, 16417–16426. [CrossRef] [PubMed]
44. Moghaddam, A.B.; Namvar, F.; Moniri, M.; Md Tahir, P.; Azizi, S.; Mohamad, R. Nanoparticles biosynthesized by fungi and yeast: A review of their preparation, properties, and medical applications. *Molecules* **2015**, *20*, 16540–16565. [CrossRef] [PubMed]
45. Pereira, L.; Mehboob, F.; Stams, A.J.M.; Mota, M.M.; Rijnaarts, H.H.M.; Alves, M.M. Metallic nanoparticles: Microbial synthesis and unique properties for biotechnological applications, bioavailability and biotransformation. *Crit. Rev. Biotechnol.* **2015**, *35*, 114–128. [CrossRef] [PubMed]
46. Chitam, H.; Zhu, N.; Shang, R.; Shi, C.; Cui, J.; Sohoo, I.; Wu, P.; Cao, Y. Biorecovery of palladium as nanoparticles by *Enterococcus faecalis* and its catalysis for chromate reduction. *Chem. Eng. J.* **2016**, *288*, 246–254.

47. Lloyd, J.R.; Byrne, J.M.; Coker, V.S. Biotechnological synthesis of functional nanomaterials. *Curr. Opin. Biotechnol.* **2011**, *22*, 509–515. [CrossRef] [PubMed]
48. Zhang, X.; Yan, S.; Tyagi, R.D.; Surampalli, R.Y. Synthesis of nanoparticles by microorganisms and their application in enhancing microbiological reaction rates. *Chemosphere* **2011**, *82*, 489–494. [CrossRef] [PubMed]
49. Huang, C.P.; Juang, C.P.; Morehart, K.; Allen, L. The removal of Cu (II) from dilute aqueous solutions by Saccharomyces cerevisiae. *Water Res.* **1990**, *24*, 433–439. [CrossRef]
50. Bhargavaa, A.; Jaina, N.; Gangopadhyayb, S.; Panwara, J. Development of gold nanoparticle-fungal hybrid based heterogeneous interface for catalytic applications. *Process Biochem.* **2015**, *50*, 1293–1300. [CrossRef]
51. Vetchinkina, E.P.; Loshchinina, E.A.; Burov, A.M.; Dykman, L.A.; Nikitina, V.E. Enzymatic formation of gold nanoparticles by submerged culture of the *basidiomycete Lentinus edodes*. *J. Biotechnol.* **2014**, *182–183*, 37–45. [CrossRef] [PubMed]
52. Mishra, A.; Kumari, M.; Pandey, S.; Chaudhry, V.; Gupta, K.C.; Nautiyal, C.S. Biocatalytic and antimicrobial activities of gold nanoparticles synthesized by *Trichoderma* sp. *Bioresour. Technol.* **2014**, *166*, 235–242. [CrossRef] [PubMed]
53. Tsai, Y.-J.; Ouyang, C.-Y.; Ma, S.-Y.; Tsai, D.-Y.; Tsengand, H.-W.; Yeh, Y.-C. Biosynthesis and display of diverse metal nanoparticles by recombinant *Escherichia coli*. *RSC Adv.* **2014**, *4*, 58717–58719. [CrossRef]
54. Mao, H.; Liao, Y.; Ma, J.; Zhao, S.L.; Huo, F.W. Water-soluble metal nanoparticles stabilized by plant polyphenols for improving the catalytic properties in oxidation of alcohols. *Nanoscale* **2016**, *8*, 1049–1054. [CrossRef] [PubMed]
55. Peshkov, V.A.; Pereshivko, O.P.; Van der Eycken, E.V. A walk around the A^3-coupling. *Chem. Soc. Rev.* **2012**, *41*, 3790–3807. [CrossRef] [PubMed]

Article

Supramolecular Assembly of Gold Nanoparticles on Carbon Nanotubes: Application to the Catalytic Oxidation of Hydroxylamines

Nimesh Shah [1,2,†], Pallabita Basu [1,†], Praveen Prakash [2,†], Simon Donck [1,2], Edmond Gravel [2], Irishi N. N. Namboothiri [1,*] and Eric Doris [2,*]

1 Department of Chemistry, Indian Institute of Technology, Bombay, Mumbai 400076, India; nimesh31@yahoo.com (N.S.); pallabita17@gmail.com (P.B.); simon.donck@gmail.com (S.D.)

2 Alternative Energies and Atomic Energy Commission (CEA), Saclay Institute of Biology and Technology (IBITECS), Department of Bioorganic Chemistry and Isotopic Labeling, 91191 Gif-sur-Yvette, France; praveen.prakash@cea.fr (P.P.); edmond.gravel@cea.fr (E.G.)

* Correspondence: irishi@chem.iitb.ac.in (I.N.N.N.); eric.doris@cea.fr (E.D.); Tel.: +91-2225767196 (I.N.N.N.); +33-169088071 (E.D.)

† These authors contributed equally to this work.

Academic Editors: Hermenegildo García and Sergio Navalón
Received: 19 January 2016; Accepted: 16 February 2016; Published: 24 February 2016

Abstract: A supramolecular heterogeneous catalyst was developed by assembly and stabilization of gold nanoparticles on the surface of carbon nanotubes. A layer-by-layer assembly strategy was used and the resulting nanohybrid was involved in the catalytic oxidation of hydroxylamines under mild conditions. The nanohybrid demonstrated high efficiency and selectivity on hydroxylamine substrates.

Keywords: carbon nanotubes; gold nanoparticles; nanohybrid; heterogeneous catalysis

1. Introduction

Besides stoichiometric approaches that are routinely used for the oxidation of organic substrates [1,2], catalytic processes have also been devised to perform selective oxidation reactions [3]. With heterogeneous catalytic systems, obtained by assembling the metallic catalysts on a solid support, facile reclaim and reuse of the catalytic species can be achieved [4]. Among the various platforms used as supports, allotrope forms of carbon, in particular carbon nanotubes (CNTs), have been shown to provide some key advantages [5]. The latter include chemical, thermal, and mechanical stability, high specific surface area, inertness, and adjustable topography. CNTs are also able to act in a synergistic fashion to enhance the performances of the supported catalytic metal [6,7]. With these features in mind, we sought to develop a catalyst that could catalyze some oxidation reactions under mild conditions of temperature and pressure using low catalytic loadings. Herein, we report the assembly of a CNT-gold nanohybrid catalyst and its application to the selective oxidation of hydroxylamines into either nitroso or azoxy derivatives.

2. Results and Discussion

2.1. *Assembly of the AuCNT Nanohybrid*

Although gold has long been regarded as a poor catalytic metal, its nanosized forms [8], including supported ones [9,10], have recently been shown to be able to catalyze a wide array of chemical transformations. Our CNT-gold catalyst was built using a layer-by-layer approach that was adapted from our previous work [11–23]. Carbon nanotubes were first dispersed in an aqueous solution by ultrasonication in the presence of an amphiphilic nitrilotriacetic-diyne (DANTA) surfactant. This first step led to the non-stochastic assembly of the amphiphilic units on the CNT surface and to the formation of supramolecular structures with a nanoring-like shape (Figure 1a,c). This type of well-ordered supramolecular assembly on CNTs was first reported in 2003 [24]. Amphiphilic DANTA adsorbed at the surface of the nanotubes by hydrophobic interactions while its hydrophilic polar head was pointing toward the aqueous medium. The rings were polymerized by ultraviolet (UV) irradiation in a second step. In fact, irradiation of the sample for 6 h at 254 nm led to a topochemical polymerization of the diyne motif incorporated in the hydrophobic part of the starting amphiphile [25,26]. Polymerization takes place within individual half-cylinders and strengthens the cohesion of the assembly. After UV irradiation, the DANTA-decorated nanotubes became resistant to dialysis against water and to ethanol washes, indicating that the lipid assemblies had been polymerized. The second layer was thereafter deposited by stirring the suspended nanotubes with a cationic polymer, poly(diallyldimethylammonium chloride) (PDADMAC), which adsorbed on the nanotube's surface by electrostatic interactions with the primary anionic layer. The double-coated CNTs were then recovered by centrifugation before the final deposition of gold nanoparticles (AuNPs). The latter were prepared in parallel by reduction of $HAuCl_4$ in the presence of tetrahydroxymethylphosphonium chloride [27]. The colloidal suspension of gold nanoparticles (AuNPs) was then sequentially added to the coated CNTs in which the polyammonium network provided robust anchoring and stabilization of the metallic nanoparticles (Figure 1b,d). The AuCNT nanohybrid was finally suspended in water and used for the catalysis of the reported reactions.

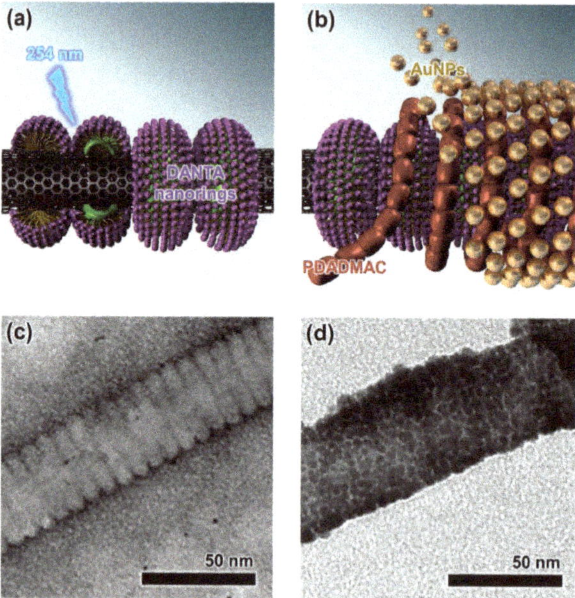

Figure 1. (**a**) Schematic representation of nitrilotriacetic-diyne (DANTA) nanoring polymerization on the carbon nanotube (CNT) surface; (**b**) Schematic representation of the final stages of the AuCNT synthesis; (**c**) Transmission electron microscopy (TEM) image of polymerized DANTA nanorings on CNT (negative staining); (**d**) TEM image of the AuCNT final catalyst.

2.2. Characterization of the AuCNT Nanohybrid

Transmission electron microscopy (TEM, Philips CM12 microscope, Amsterdam, the Netherlands) indicated that the AuNPs were of spherical shape. The average size of the supported nanoparticles was measured using TEM pictures. Statistical measurement indicated that the supported AuNPs had an average diameter of *ca.* 3 nm (Figure 2a). The metal content of the aqueous AuCNT suspension was determined by inductively coupled plasma mass spectrometry (ICP-MS) ([Au] = 1.2 mM). Finally, the metallic character of the supported gold nanoparticles was established by X-ray photoelectron spectroscopy (XPS, VG ESCALAB 210 spectrometer, Waltham, MA, USA) analysis which showed characteristic Au 4f-binding energies contributions of Au [28] (Figure 2b).

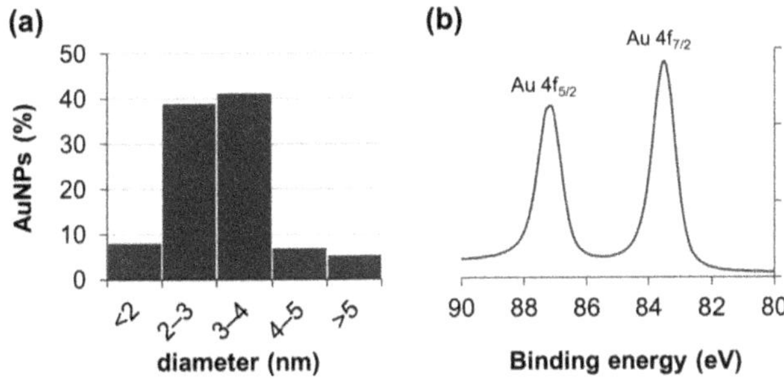

Figure 2. (**a**) Size distribution obtained from the measurement of 250 gold particles; (**b**) X-ray photoelectron spectroscopy (XPS) analysis (Au 4f core level) of the AuCNT nanohybrid.

2.3. *Oxidation of Hydroxylamines with the AuCNT Nanohybrid*

With the gold-based nanohybrid in hand, we investigated its potential in the aerobic oxidation of hydroxylamines. In the latter transformation, we expected the formation of the corresponding nitroso derivatives [29]. The nanohybrid-catalyzed oxidation of *tert*-butyl hydroxylamine (**1a**) in $CHCl_3/H_2O$ afforded *tert*-butyl nitroso compound **1b** in 81% yield after 12 h of reaction (Table 1, Entry 1). The use of a binary solvent mixture permitted us to increase the rate of the reaction and afforded a more selective transformation. The reaction of *N*-cyclohexylhydroxylamine (**2a**) also cleanly produced nitrosocyclohexane (**2b**) in 83% yield (Entry 2). It is noteworthy that no isomerization of nitrosocyclohexane (**2b**) into the corresponding cyclohexanone oxime was detected. This result is to be noted since tautomerization of nitroso compounds to oximes is classically observed when working with substrates carrying Cα-protons. The reaction of aliphatic substrates thus satisfactorily provided access to nitroso compounds in high yields, but aromatic hydroxylamines did not behave similarly.

Table 1. Oxidation of various hydroxylamines with AuCNT [a].

R–NH–OH (**a**) —AuCNT→ oxidized product **b**

Entry	Substrate a	Product b	Yield (%)
1	*tert*-butyl–NH–OH	*tert*-butyl–N=O	81 [b]
2	cyclohexyl–NH–OH	cyclohexyl–N=O	83 [b]

Table 1. *Cont.*

Entry	Substrate a	Product b	Yield (%)
3	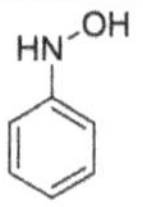	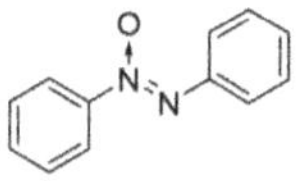	96 [c]
4	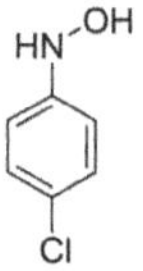	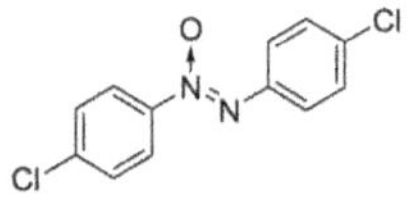	98 [c]
5	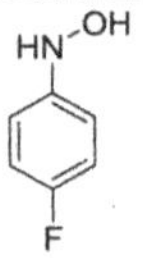	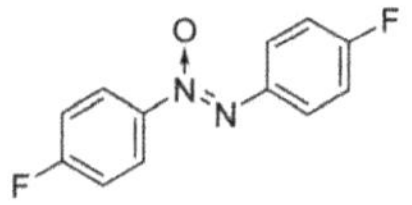	92 [c]
6	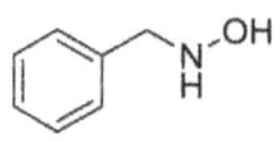	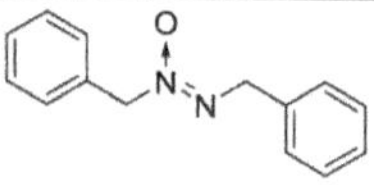	94 [c]
7	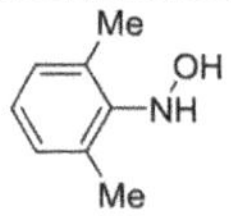	NR	– [c]

[a] Conditions: *N*-hydroxylamine (0.1 mmol), AuCNT (0.5 mol %), K_2CO_3 (0.2 mmol), room temperature under air. [b] Reaction time 12 h, $CHCl_3/H_2O$ (1:1, 2 mL). [c] Reaction time 2 h, $CHCl_3$ (2 mL).

For example, the reaction of phenylhydroxylamine (**3a**) with AuCNT in $CHCl_3$ gave a dimeric product in the form of the azoxy derivative **3b** in 2 h only (Entry 3). It is noteworthy that the binary solvent system was not required in the case of aromatic hydroxylamine substrates. A possible mechanism for the conversion observed in entry 3 (formation of the azoxy compound) could involve the oxidation of phenylhydroxylamine (**3a**) into the corresponding nitroso derivative **3a′** (Scheme 1). As **3a′** accumulates in the solution, its condensation with the unreacted phenylhydroxylamine (**3a**) leads to the formation of azoxy derivative **3b**, after the elimination of a molecule of water.

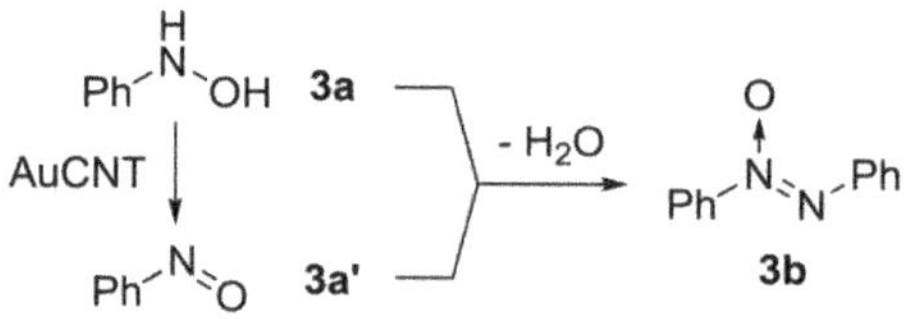

Scheme 1. Postulated mechanism for the formation of azoxy compounds.

The same type of reactivity was also evidenced for other aromatic substrates such as 4-chlorophenylhydroxylamine (**4a**, Entry 4) and 4-fluorophenylhydroxylamine (**5a**, Entry 5). Indeed, we detected the formation of azoxy dimers **4b** (98% yield) and **5b** (92% yield) starting from compounds **4a** and **5a**, respectively. The transformation worked equally well on benzylhydroxylamine (**6a**) in 94% yield (Entry 6) but failed on the more hindered 2,6-dimethylphenylhydroxylamine (**7a**, Entry 7) which remained unaffected.

2.3.1. Recycling of the AuCNT Nanohybrid

To assess the recyclability of the AuCNT nanohybrid, multiple oxidation cycles were carried out by successive reuse of the same sample of catalyst. A classical oxidation reaction was set using the general procedure described above which was applied to *N*-phenylhydroxylamine (**3a**). After completion, the catalyst was recovered by centrifugation, and the supernatant was worked up. The catalyst was washed with tetrahydrofuran (THF) and reused in subsequent oxidation reactions. This process was repeated over five consecutive cycles and showed no significant decrease in yields of oxidized azo product **3b** (Table 2). After the fifth run, TEM analysis showed no major alteration of the nanohybrid morphology. The use of non-supported AuNPs provided lower yields of product and the catalyst could not be recycled.

Table 2. Recycling of AuCNT for the oxidation of *N*-phenylhydroxylamine.

3a Ph–NH–OH —AuCNT (0.5 mol%), $CHCl_3$, 2 h→ Ph–N=N(→O)–Ph **3b**

Entry	AuCNT	Yield (%)
1	fresh	96
2	1st reuse	93
3	2nd reuse	91
4	3rd reuse	93
5	4th reuse	94

2.3.2. Is AuCNT a Heterogeneous Catalyst?

The involvement of the nanohybrid in the oxidation process was demonstrated by the following experiment. A standard oxidation reaction of *N*-phenylhydroxylamine (**3a**) was run and, after 45 min, split into two. At this stage, approximately 50% conversion of **3a** into the corresponding azoxy derivative **3b** was detected by ^{1}H-NMR (Bruker Avance DPX 400 MHz spectrometer, Billerica, MA, USA). The AuCNT catalyst was removed by centrifugation in one sample, whereas it was left in the other. Both reactions were further stirred for additional time before analysis. While the reaction was nearly quantitative (*ca.* 95% yield) in the AuCNT-containing sample, no further conversion was detected in the absence of the nanohybrid. These observations confirmed that oxidation of *N*-phenylhydroxylamine (**3a**) occurred at the surface of the AuCNT nanohybrid which acted as a solid catalyst [30] and whose performances are comparable to that of a previously developed rhodium-carbon nanotube catalytic system [23].

3. Experimental Section

3.1. Assembly of the Nanohybrid

Amphiphilic DANTA (20 mg) was dissolved in 2 mL of 25 mM pH 8 aqueous Tris-buffer before multiwalled carbon nanotubes (50 mg) were added. The dispersion was sonicated and the stable suspension transferred into two tubes. The tubes were centrifuged at 5000× *g* and the supernatants were collected. The latter were centrifuged at 15,000× *g* for 45 min. The supernatant was discarded and the pellets taken in buffer and centrifuged again at 15,000× *g* for 45 min. The pellets were finally resuspended in 1.5 mL of buffer and submitted to UV irradiation at 254 nm for 6 h.

After polymerization, the buffer volume was adjusted to 1.5 mL. The suspension was stirred in the presence of the cationic polymer PDADMAC (700 μL of a 20% water solution) for 1 h. The ensuing centrifugation at 15,000× *g* for 30 min permitted to get rid of the polymer in excess. The pellets were taken in 2 mL of buffer. This operation was repeated twice using the buffer solution and two more times using pure water.

The final pellets were resuspended in 1 mL of water. Then 50 μL of the latter suspension was transferred to Eppendorf® tubes (×20). To each tube was added 1 mL of a 1 mM colloid suspension of the gold nanoparticles [27] and the mixture was vortex-stirred at room temperature for 1 min every 30 min (during 4 h). The suspension was then centrifuged at 3000× *g* for 5 min. The supernatant was discarded and 1 mL of a fresh gold colloid suspension was added. The same process was repeated two more times. The pellets were washed three times by centrifugation/redispersion in water. The 20 pellets were combined and 4 mL of water was finally added.

3.2. Procedure for the Oxidation of Hydroxylamines

A typical procedure is given for the oxidation of *N*-cyclohexylhydroxylamine **2a**. Under air, to a stirred solution of *N*-cyclohexylhydroxylamine hydrochloride (**2a**.HCl, 0.1 mmol) in 2 mL of $CHCl_3/H_2O$ (1:1) was added K_2CO_3 (2 equivalents) and 0.5 mol % of the suspension of the AuCNT catalyst. The reaction mixture was stirred at room temperature for 12 h. The aqueous layer was extracted with $CHCl_3$. The combined organic layer was dried over anhydrous Na_2SO_4, filtered, and concentrated under vacuum. The crude residue was purified by column chromatography to afford nitroso-cyclohexyl amine **2b** in 83% yield. ^{1}H-NMR (400 MHz, $CDCl_3$) δ (ppm) 5.07 (triplet of triplet, J = 3.8 Hz, J = 11.6 Hz, 1H), 1.97–1.94 (multiplet, 2H), 1.89–1.86 (multiplet, 2H), 1.69–1.65 (multiplet, 2H), 1.43–1.31 (multiplet, 2H), 1.28–1.20 (multiplet, 2H); ^{13}C-NMR (100 MHz, $CDCl_3$) δ (ppm) 65.6, 28.2 (2C), 25.0, 24.5 (2C).

4. Conclusions

A nanohybrid catalyst was produced by the layer-by-layer supramolecular assembly of gold nanoparticles on carbon nanotubes. The gold-based nanohybrid (AuCNT) was employed in the oxidation of hydroxylamines to provide straightforward access to the corresponding oxidized products in good to excellent yields. The transformation led either to nitroso derivatives in the case of aliphatic hydroxylamines or azoxy derivatives in the case of aromatic/benzylic hydroxylamines. Selectivity, low catalyst loading and mild reaction conditions (e.g., room temperature, open air) are the salient features of the AuCNT methodology.

Acknowledgments: Supports from the Indo-French Centre for the Promotion of Advanced Research (IFCPAR)/Centre Franco-Indien pour la Promotion de la Recherche Avancée (CEFIPRA) (Project No. 4705-1) and from the CEA-Enhanced Eurotalents program are gratefully acknowledged. The TEM-team platform (CEA, IBITECS) is acknowledged for help with TEM images. The "Service de Chimie Bioorganique et de Marquage" belongs to the Laboratory of Excellence in Research on Medication and Innovative Therapeutics (ANR-10-LABX-0033-LERMIT).

Author Contributions: N.S., P.B., P.P. and S.D. performed the experiments. E.G., I.N.N.N. and E.D. designed the experiments and wrote the manuscript.

Conflicts of Interest: The authors declare no conflict of interest.

References

1. Donohoe, T.J. *Oxidation and Reduction in Organic Synthesis*; Oxford University Press: Oxford, UK, 2000.
2. Hussain, H.; Green, I.R.; Ahmed, I. Journey describing applications of oxone in synthetic chemistry. *Chem. Rev.* **2013**, *113*, 3329–3371. [CrossRef] [PubMed]
3. Van Santen, R.A. *Catalytic Oxidation: Principles and Applications*; Sheldon, R.A., van Santen, R.A., Eds.; World Scientific: Singapore, 1995.
4. Ali, M.E.; Shaheen, M.M.; Sarkar, M.; Hamid, S.B.A. Heterogeneous metal catalysts for oxidation reactions. *J. Nanomater.* **2014**, *2014*. [CrossRef]
5. John, J.; Gravel, E.; Namboothiri, I.N.N.; Doris, E. Advances in carbon nanotube-noble metal catalyzed organic transformations. *Nanotechnol. Rev.* **2012**, *1*, 515–539. [CrossRef]
6. Singh, R.; Premkumar, T.; Shin, J.Y.; Geckeler, K.E. Carbon nanotube and gold-based materials: A symbiosis. *Chem. Eur. J.* **2010**, *16*, 1728–1743. [CrossRef] [PubMed]

7. Melchionna, M.; Marchesan, S.; Prato, M.; Fornasiero, P. Carbon nanotubes and catalysis: The many facets of a successful marriage. *Catal. Sci. Technol.* **2015**, *5*, 3859–3875. [CrossRef]
8. Zhang, Y.; Cui, X.; Shi, F.; Deng, Y. Nano-gold catalysis in fine chemical synthesis. *Chem. Rev.* **2012**, *112*, 2467–2505. [CrossRef] [PubMed]
9. Gravel, E.; Namboothiri, I.N.N.; Doris, E. Supramolecular assembly of gold nanoparticles on carbon nanotubes and catalysis of selected organic transformations. *Synlett* **2016**, *27*. [CrossRef]
10. Stratakis, M.; Garcia, H. Catalysis by supported gold nanoparticles: Beyond aerobic oxidative processes. *Chem. Rev.* **2012**, *112*, 4469–4506. [CrossRef] [PubMed]
11. John, J.; Gravel, E.; Hagège, A.; Li, H.; Gacoin, T.; Doris, E. Catalytic oxidation of silanes by carbon nanotube-gold nanohybrid. *Angew. Chem. Int. Ed.* **2011**, *50*, 7533–7536. [CrossRef] [PubMed]
12. Kumar, R.; Gravel, E.; Hagège, A.; Li, H.; Jawale, D.V.; Verma, D.; Namboothiri, I.N.N.; Doris, E. Carbon nanotube-gold nanohybrids for selective catalytic oxidation of alcohols. *Nanoscale* **2013**, *5*, 6491–6497. [CrossRef] [PubMed]
13. Kumar, R.; Gravel, E.; Hagège, A.; Li, H.; Verma, D.; Namboothiri, I.N.N.; Doris, E. Direct reductive amination of aldehydes catalyzed by CNT-gold nanohybrids. *ChemCatChem* **2013**, *5*, 3571–3575. [CrossRef]
14. Jawale, D.V.; Gravel, E.; Geertsen, V.; Li, H.; Shah, N.; Namboothiri, I.N.N.; Doris, E. Aerobic oxidation of phenols and related compounds using carbon nanotube-gold nanohybrid catalysts. *ChemCatChem* **2014**, *6*, 719–723. [CrossRef]
15. Jawale, D.V.; Gravel, E.; Geersten, V.; Li, H.; Shah, N.; Namboothiri, I.N.N.; Doris, E. Size effect of gold nanoparticles supported on carbon nanotube as catalysts in selected organic reactions. *Tetrahedron* **2014**, *70*, 6140–6145. [CrossRef]
16. Shah, N.; Gravel, E.; Jawale, D.V.; Doris, E.; Namboothiri, I.N.N. Carbon nanotube-gold nanohybrid catalyzed *N*-formylation of amines using aqueous formaldehyde. *ChemCatChem* **2014**, *6*, 2201–2205. [CrossRef]
17. Shah, N.; Gravel, E.; Jawale, D.V.; Doris, E.; Namboothiri, I.N.N. Synthesis of quinoxalines via carbon nanotube-gold nanohybrid catalyzed cascade reaction of vicinal diols and ketoalcohols with diamines. *ChemCatChem* **2015**, *7*, 57–61. [CrossRef]
18. Jawale, D.V.; Gravel, E.; Boudet, C.; Shah, N.; Geertsen, V.; Li, H.; Namboothiri, I.N.N.; Doris, E. Selective conversion of nitroarenes using a carbon nanotube-ruthenium nanohybrid. *Chem. Commun.* **2015**, *51*, 1739–1742. [CrossRef] [PubMed]
19. Jawale, D.V.; Gravel, E.; Boudet, C.; Shah, N.; Geertsen, V.; Li, H.; Namboothiri, I.N.N.; Doris, E. Room temperature Suzuki coupling of aryl iodides, bromides, and chlorides using a heterogeneous carbon nanotube-palladium nanohybrid catalyst. *Catal. Sci. Technol.* **2015**, *5*, 2388–2392. [CrossRef]
20. Jawale, D.V.; Gravel, E.; Shah, N.; Dauvois, V.; Li, H.; Namboothiri, I.N.N.; Doris, E. Cooperative dehydrogenation of *N*-heterocycles using a carbon nanotube-rhodium nanohybrid. *Chem. Eur. J.* **2015**, *21*, 7039–7042.
21. Donck, S.; Gravel, E.; Shah, N.; Jawale, D.V.; Doris, E.; Namboothiri, I.N.N. Tsuji-Wacker oxidation of terminal olefins using a palladium-carbon nanotube nanohybrid. *ChemCatChem* **2015**, *7*, 2318–2322. [CrossRef]
22. Donck, S.; Gravel, E.; Shah, N.; Jawale, D.V.; Doris, E.; Namboothiri, I.N.N. Deoxygenation of amine *N*-oxides using gold nanoparticles supported on carbon nanotubes. *RSC Adv.* **2015**, *5*, 50865–50868. [CrossRef]
23. Donck, S.; Gravel, E.; Li, A.; Prakash, P.; Shah, N.; Leroy, J.; Namboothiri, I.N.N.; Doris, E. Mild and selective catalytic oxidation of organic substrates by a carbon nanotube-rhodium nanohybrid. *Catal. Sci. Technol.* **2015**, *5*, 4542–4546. [CrossRef]
24. Richard, C.; Balavoine, F.; Schultz, P.; Ebbesen, T.W.; Mioskowski, C. Supramolecular self-assembly of lipid derivatives on carbon nanotubes. *Science* **2003**, *300*, 775–778. [CrossRef] [PubMed]
25. Ogawa, K.; Tamura, H.; Hatada, M.; Ishihara, T. Study of photoreaction processes of PDA Langmuir films. *Langmuir* **1988**, *4*, 903–906. [CrossRef]
26. Shimizu, T.; Masuda, M.; Minamikawa, H. Supramolecular nanotube architectures based on amphiphilic molecules. *Chem. Rev.* **2005**, *105*, 1401–1443. [CrossRef] [PubMed]
27. Duff, D.G.; Baiker, A.; Edwards, P.P. A new hydrosol of gold clusters. Formation and particle size variation. *Langmuir* **1993**, *9*, 2301–2309. [CrossRef]
28. Conte, M.; Hutchings, G.H. *Modern Gold Catalyzed Synthesis*; Hashmi, A.S.K., Toste, F.D., Eds.; Wiley-VCH: Weinheim, Germany, 2012.

29. Gowenlock, B.G.; Richter-Addo, G.B. Preparations of C-nitroso compounds. *Chem. Rev.* **2004**, *104*, 3315–3340. [CrossRef] [PubMed]
30. Astruc, D.; Lu, F.; Aranzaes, J.R. Nanoparticles as recyclable catalysts: The frontier between homogeneous and heterogeneous catalysis. *Angew. Chem. Int. Ed.* **2005**, *44*, 7852–7872. [CrossRef] [PubMed]

 *nanomaterials* 

Article

Coupling of Nanocrystalline Anatase TiO_2 to Porous Nanosized $LaFeO_3$ for Efficient Visible-Light Photocatalytic Degradation of Pollutants

Muhammad Humayun, Zhijun Li, Liqun Sun, Xuliang Zhang, Fazal Raziq, Amir Zada, Yang Qu * and Liqiang Jing *

Key Laboratory of Functional Inorganic Materials Chemistry, Ministry of Education, School of Chemistry and Materials Science, Heilongjiang University, Harbin 150080, China; humayun096@yahoo.com (M.H.); ushlj2008@163.com (Z.L.); sunliqun001@163.com (L.S.); zxlzs007@gmail.com (X.Z.); Rabiabi73@gmail.com (F.R.); amistry009@yahoo.com (A.Z.)

* Correspondence: copy0124@126.com (Y.Q.); jinglq@hlju.edu.cn (L.J.); Tel./Fax: +86-451-8660-4760 (Y.Q. and L.J.)

Academic Editors: Hermenegildo García and Sergio Navalón
Received: 26 November 2015; Accepted: 15 December 2015; Published: 20 January 2016

Abstract: In this work we have successfully fabricated nanocrystalline anatase TiO_2/perovskite-type porous nanosized $LaFeO_3$ (T/P-LFO) nanocomposites using a simple wet chemical method. It is clearly demonstrated by means of atmosphere-controlled steady-state surface photovoltage spectroscopy (SPS) responses, photoluminescence spectra, and fluorescence spectra related to the formed OH^- radical amount that the photogenerated charge carriers in the resultant T/P-LFO nanocomposites with a proper mole ratio percentage of TiO_2 display much higher separation in comparison to the P-LFO alone. This is highly responsible for the improved visible-light activities of T/P-LFO nanocomposites for photocatalytic degradation of gas-phase acetaldehyde and liquid-phase phenol. This work will provide a feasible route to synthesize visible-light responsive nano-photocatalysts for efficient solar energy utilization.

Keywords: nanostructures; semiconductors; chemical synthesis; X-ray diffraction; catalytic properties

1. Introduction

Photocatalytic materials have received tremendous attention in recent years due to the increase in world-wide environmental pollution. Numerous photocatalysts including nanoparticles of oxides, noble metals, and their nanocomposites have been explored for their superior performance in the degradation of organic pollutants under visible irradiation [1,2]. Recently, considerable attention has been focused on perovskite-type oxides with the general formula ABO_3, where site A is a rare-earth element and site B is 3D transition metal [3]. In perovskite oxides, the presence of B-site metal cations and oxygen vacancies are of great significance, because the catalytic process mainly depends on the redox properties of B-site metal cations, whereas the oxygen vacancies provide the activation and adsorption sites for the substrates [4]. Hence, it is easy to alter the energy band gap, photogenerated charge separation and then photocatalytic activity [5,6]. It is generally accepted that there is a great potential for ABO_3-type oxides to be taken as efficient photocatalysts. Among the well-known ABO_3-type perovskite oxides, $LaFeO_3$ (LFO) has attracted much attention owing to its potential applications in photocatalysis, gas sensors, solid-oxide fuel cells, and electronic and magnetic materials [7].

LFO has been chosen as an efficient photocatalyst due its narrow band gap (2.0 eV), which is active under visible light [8]. However, the photocatalytic activity of LFO is still limited, which is attributed to the weak photogenerated charge separation. Similar to other visible-responsive oxide photocatalysts, it has a low conduction band position, located below the standard hydrogen electrode reduction level, which allows fast recombination of photogenerated charge-carriers [9]. Another drawback of traditional LFO is that the materials possess low surface areas [10]. To improve the photocatalytic performance of LFO, elemental doping, coupling with semiconducting metal-oxides and increasing surface areas by introducing pores are widely employed [4,11–13]. In general, it is widely accepted that the photogenerated high-energy electrons of narrow band-gap oxides could relax to the bottom of conduction band in extremely short time [2]. This would lead to the fast recombination of photogenerated charges. To overcome this shortfall, the couplings of wide band gap oxides are highly desirable.

In our previous works [2,14], the visible-light activities of $BiVO_4$ and Fe_2O_3 were obviously improved by coupling TiO_2 with high-level conduction bottom, primarily demonstrating that this idea is feasible. In such heterojunctions, the visible-light-excited high-energy electrons (from valance band (VB) of narrow band gap oxides) could transfer thermodynamically from the CB of narrow band gap oxides to the other constitution with high energy platform. To the best of our knowledge, there is no previous report on the enhanced visible-light activities of large surface area porous LFO for photocatalytic degradation of colorless organic pollutants by coupling with nanocrystalline anatase TiO_2.

Herein, we present our work on the visible-light enhanced photocatalytic activities of T/P-LFO nanocomposites for efficient degradation of gas-phase acetaldehyde and liquid-phase phenol. Based on our experimental results, it is suggested that the enhanced photoactivities are attributed to the improved separation of electron-hole pairs in the resulting T/P-LFO nanocomposites. This work will provide feasible routes to synthesize visible-light responsive nano-photocatalysts for efficient solar energy utilization.

2. Results and Discussion

2.1. Structural Characterization and Surface Composition

Figure 1A shows the X-ray diffraction (XRD) patterns of P-LFO, TiO_2 (anatase) and T/P-LFO nanocomposites. The XRD patterns of P-LFO calcined at 600 °C show diffraction peaks at 22.45, 32.13, 39.66, 46.06, 52.01, 57.42, 67.32, 76.63, respectively, correspond to the (101), (121), (220), (202), (141), (240), (242) and (204) reflections of orthorhombic phase $LaFeO_3$ (JCPDS No. 37-1493) without any impurity phase. The sharp peaks imply the high crystallinity of the product [15]. In the XRD patterns of T/P-LFO nanocomposites calcined at 450 °C, no impurity peaks can be observed except for TiO_2 (T) and LFO phases. In addition, the relative peak intensity of TiO_2 in T/P-LFO nanocomposites is obviously enhanced with an increase in mole ratio percentage of coupled TiO_2.

The particle size of P-LFO was calculated using the Scherrer formula, which is approximately 50 nm. According to the UV-vis DRS spectra (Figure 1B), the absorption lower than 400 nm for TiO_2 is attributed to the electronic transitions from its valance band to the conduction band (O2p → La3d). While the absorption edge of P-LFO appeared at approximately 620 nm, attributed to its electronic transition from the valance band to conduction band (O2p → Fe3d), and the optical absorption across the energy band gap is 2.0 eV.

To investigate the morphology and microstructure of P-LFO and 9T/P-LFO nanocomposite, the TEM and HRTEM micrographs were taken. One can see that the as-synthesized P-LFO exhibits random distribution with a crystallite size of about 50 nm determined from the TEM micrograph (Figure 1C). This is consistent with the crystallite size calculated from the XRD patterns. Moreover, the nanocrystalline orthorhombic-phase structure of P-LFO is confirmed from the selected area electron diffraction (SAED) patterns (inset of Figure 1C). From the HRTEM micrograph of the 9T/P-LFO nanocomposite (Figure 1D inset), it can be seen that an intimate interface junction exists between TiO_2 and P-LFO. The lattice fringes at (121) plane with d-spacing 0.28 nm correspond to the orthorhombic phase LFO [16], while the latter at (101) plane with d-spacing 0.35 nm are attributed to the anatase TiO_2 [17].

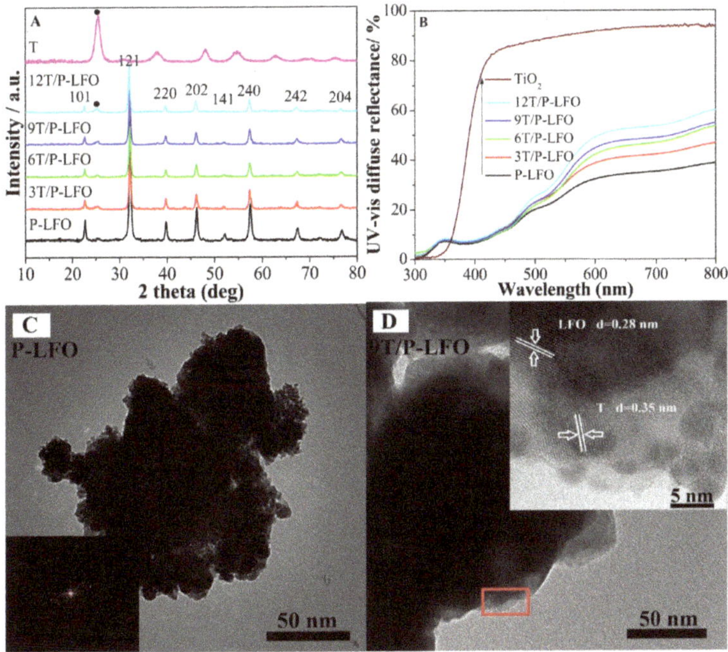

Figure 1. X-ray diffraction (XRD) patterns (**A**) and UV-vis diffuse reflectance (UV-vis DRS) spectra (**B**) of Porous-$LaFeO_3$ (P-LFO) TiO_2 (T) and TiO_2/P-LFO (T/P-LFO) nanocomposites; Transmission electron microscopy (TEM) image of P-LFO with inset selected area electron diffraction (SAED) pattern (**C**); TEM image of 9T/P-LFO with inset high resolution transmission electron microscopy (HRTEM) image (**D**).

Figure 2A shows the SEM micrograph of P-LFO calcined at 600 °C. It can be seen clearly, that the P-LFO exhibits an irregular porous morphology. Moreover, it is suggested that the P-LFO nanoparticles are inter-connected and a large network of irregular shapes and sizes are formed due to the escape of a large number of gases due to the strong redox reaction taking place during the sol-gel auto-combustion. After coupling with nanocrystalline anatase TiO_2, the morphology of the nanoparticles are slightly changed as shown in Figure 2B. The XRD, TEM and SEM results demonstrate that TiO_2/P-LFO nanocomposites were successfully fabricated.

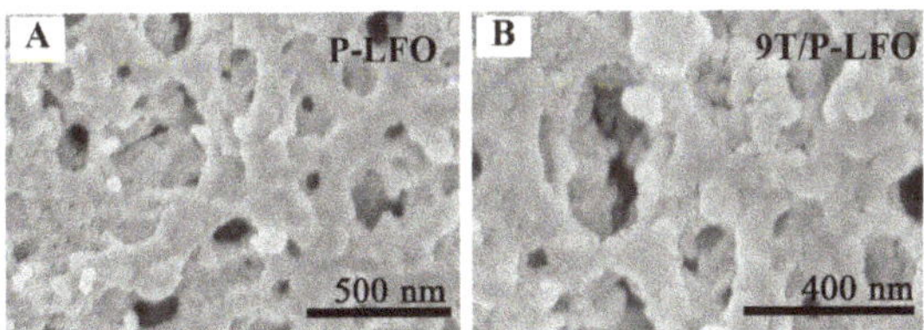

Figure 2. Scanning electron microscopy (SEM) micrograph of P-LFO (**A**) and 9T/P-LFO nanocomposite (**B**).

The surface functional groups and elemental states of the P-LFO and 9T/P-LFO nanocomposite was characterized by X-ray photoelectron spectroscopy (XPS). The binding energies were calibrated with respect to the adventitious carbon (C1s) as a reference line at 284.6 eV. The typical survey spectra (Figure 3A), of P-LFO and 9T/P-LFO samples reveals the presence of La3d, Fe2p, O1s, Ti2p and C1s. In the high-resolution spectrum of La3d for P-LFO (Figure 3B), two intense peaks, observed at 833.6

and 850.5 eV respectively, correspond to the spin-orbital splitting of $3d_{5/2}$ and $3d_{3/2}$ of La^{3+} ions in the oxide form. It can be observed that after coupling with TiO_2, the binding energies of La3d for 9T/P-LFO nanocomposite are slightly shifted toward lower binding energies. For Fe2p (Figure 3C), the binding energies observed at 710 ($Fe2p_{3/2}$) and 723.6 eV ($Fe2p_{1/2}$) respectively, correspond to the +3 oxidation state of Fe in P-LFO oxide. In addition, the binding energies of Fe2p for 9T/P-LFO nanocomposite are also shifted toward lower energies. The XPS spectra of O1s (Figure 3D) are fitted into two separate peaks with origin software by the Gaussian rule. The XPS spectra of O1s are broad and asymmetric, demonstrating that there exist two kinds of O chemical states, including crystal lattice oxygen (OL) and hydroxyl oxygen (OH) with increasing binding energy. The XPS signal for OL corresponds to the La-O and Fe-O in P-LFO crystal lattice and appeared at approximately 529.5 eV, while the OH XPS signal lies at about 531.5 eV, and is closely related to the hydroxyl species resulting from the chemisorbed water [18,19]. Besides, the binding energy of Ti2p (Figure 3E) at 458 eV corresponds to TiO_2 in 9T/P-LFO nanocomposite [20]. Hence, it is suggested that an intimate contact exist between TiO_2 and P-LFO interfaces. Obviously, this is in good agreement with the TEM results.

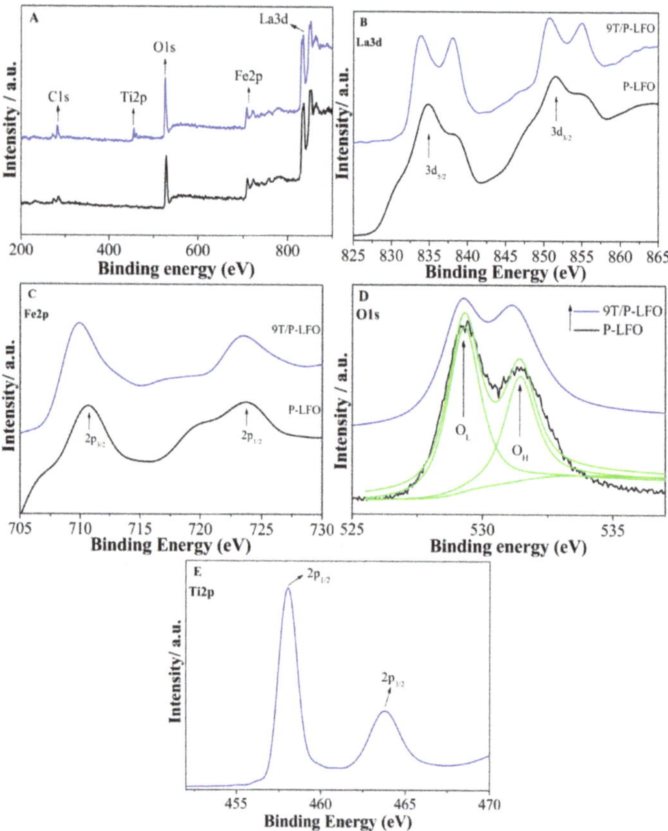

Figure 3. X-ray photoelectron spectroscopy (XPS) survey spectra of P-LFO and 9T/P-LFO nanocomposite (**A**); with high resolution images La3d (**B**); Fe2p (**C**); O1s (**D**); Ti2p (**E**).

The nitrogen adsorption-desorption isotherms and the corresponding pore size distribution of P-LFO and 9T/P-LFO nanocomposite respectively, are depicted in Figure 4A,B. It can be seen clearly that the sorption isotherms of P-LFO and 9T/P-LFO exhibit hysteresis loops, which are the characteristics of porous structure [16]. The Brunauer-Emmett-Teller (BET) surface area observed for P-LFO is 23.2 $m^2 \cdot g^{-1}$ and its BJH average pore diameter is 7.4585 nm, while for 9T/P-LFO nanocomposite, the BET surface area is considerably higher (34.2 $m^2 \cdot g^{-1}$) as compared to P-LFO. This is well attributed to the small size with large surface area of coupled TiO_2 [21].

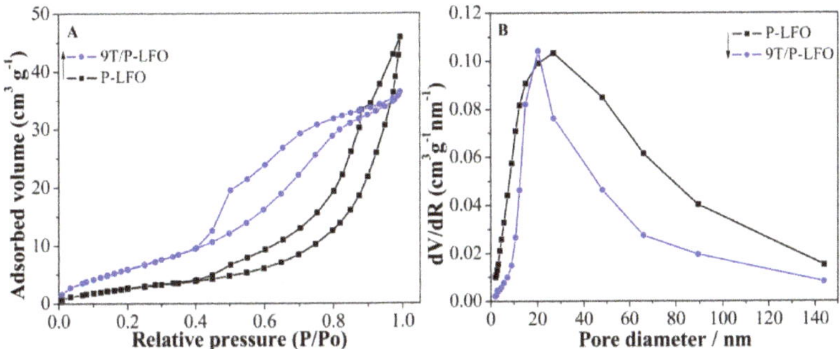

Figure 4. N_2 adsorption/desorption isotherms (**A**) and pore diameter (**B**) of P-LFO and 9T/P-LFO nanocomposite.

2.2. Photogenerated Charge Properties

Surface photovoltage spectroscopy (SPS) is a highly sensitive and non-destructive technique used to study the photophysics of the photogenerated charges in semiconducting solid materials, resulting from the changes of surface potential barriers before and after illumination. The SPS response for nanocrystalline semiconductors would mainly be derived from photo-generated charge separation via the diffusion process. From the SPS responses of P-LFO in different atmospheres in Figure 5A, it can be seen that the SPS response intensity is greatly influenced by the amount of O_2. In N_2 atmosphere, no obvious SPS response is detected for P-LFO, suggesting that the presence of O_2 is necessary for SPS response to occur. This supports the role of adsorbed O_2 in capturing the photogenerated electrons. However, for 9T/P-LFO nanocomposite Figure 5B, a remarkable SPS response is detected in N_2 atmosphere, although its SPS response is also enhanced with the increase in O_2 concentration. This unexpected SPS response in N_2 is mainly attributed to the improved separation of photogenerated charges in the fabricated T/P-LFO nanocomposites [2]. From Figure 5C, it can be seen clearly that the SPS response of P-LFO in air atmosphere is obviously increased after coupling with TiO_2 and the strongest response is observed for 9T/P-LFO nanocomposite, suggesting that the charge transfer and separation is significantly improved [9,22]. However, the excess amount of TiO_2 is unfavorable for the charge transfer and separation. From SPS measurements, it is confirmed that T/P-LFO nanocomposites could exhibit high photoactivities.

The photoluminescence (PL) is a highly sensitive and non-destructive technique widely used to investigate the structure and properties of active sites on the surface of metal oxides. It always gives us information about the surface defects and oxygen vacancies, as well as about charge carrier trapping, immigration and transfer [22]. The PL spectra of P-LFO and T/P-LFO nanocomposites under excitation wavelength of 325 nm are depicted in (Figure 5D). It can be observed that the PL peak intensity of P-LFO is considerably decreased after coupling with nanocrystalline TiO_2, and the lowest PL response is detected for 9T/P-LFO nanocomposite, suggesting that the charge recombination is significantly reduced. These results are in good agreement with the above SPS results.

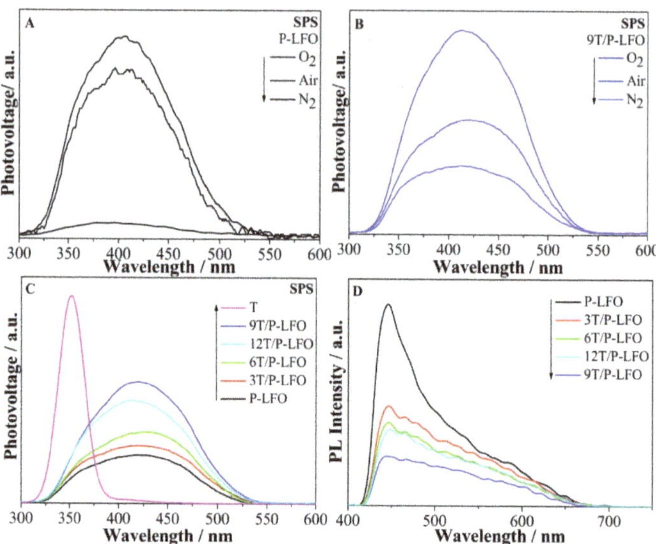

Figure 5. Surface photovoltage spectroscopy (SPS) responses of P-LFO (**A**) and 9T/P-LFO (**B**) in different atmospheres; SPS responses of P-LFO and T/P-LFO nanocomposites in air (**C**); Photoluminescence (PL) responses of P-LFO and T/P-LFO nanocomposites (**D**).

2.3. Visible-Light Photoactivities

To evaluate the visible-light activities of P-LFO and T/P-LFO nanocomposites for pollutant degradation, gas-phase acetaldehyde and liquid-phase phenol were chosen as model pollutants. The visible-light photocatalytic activity of P-LFO and T/P-LFO nanocomposites for acetaldehyde and phenol degradation is depicted in Figure 6A. It can clearly be seen that the photocatalytic degradation rates of P-LFO for acetaldehyde and phenol are greatly enhanced after coupling with TiO_2. Interestingly, 9T/P-LFO exhibits the highest photocatalytic activity. Hence, it is suggested that coupling a proper mole ratio percentage of TiO_2 is favorable for charge transfer and separation, which further leads to the enhanced visible-light activities [14].

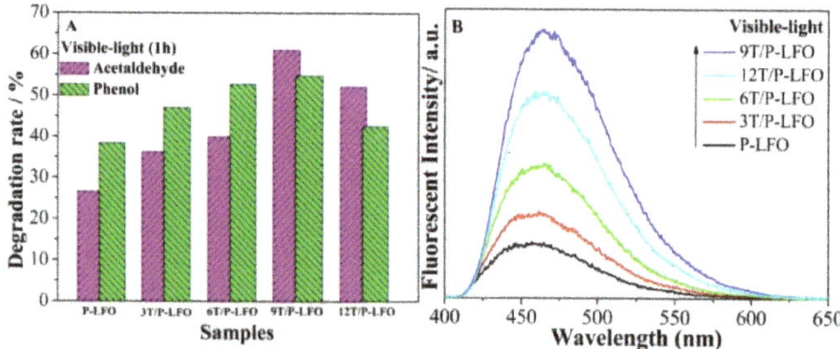

Figure 6. Visible-light photocatalytic activity for acetaldehyde and phenol degradation (**A**) and OH^- radical amount related Fluorescence spectra (**B**) of P-LFO and T/P-LFO nanocomposites.

2.4. Discussion

To prove the visible-light enhanced charge transfer and separation in the fabricated T/P-LFO nanocomposites, the coumarin fluorescent method was used to detect the amount of formed (•OH) species. As illustrated in our previous reported work [23], the amount of (•OH) species could also effectively reveal the separation of photogenerated charges in photocatalysis. It is demonstrated that in coumarin fluorescent process, the coumarin could easily react with the formed •OH species and produce luminescent 7-hydroxy-coumarin. From (Figure 6B), it can be observed that the fluorescent response intensity of P-LFO is obviously enhanced after coupling with TiO_2 and 9T/P-LFO nanocomposite exhibit the strongest fluorescent intensity peak. This further supports the SPS results and photoactivities.

Based on the above results and discussion, a possible mechanism for charge transfer and separation in T/P-LFO nanocomposites is proposed as depicted in Figure 7. It is suggested that when T/P-LFO nanocomposite is radiated by visible-light with photon energy higher than the band gap of P-LFO, electron-hole pairs are produced. The high-energy electrons would transfer thermodynamically to the CB of TiO_2, which would probably react with the surface adsorbed O_2 to produce superoxide radicals $\bullet O_2^-$, while the holes will remain in the VB of P-LFO and react with the surface adsorbed water or OH^- to produce ·OH radicals. In this way the photogenerated charge recombination could be effectively reduced and utilized in the photocatalytic process to enhance the visible-light photoactivities.

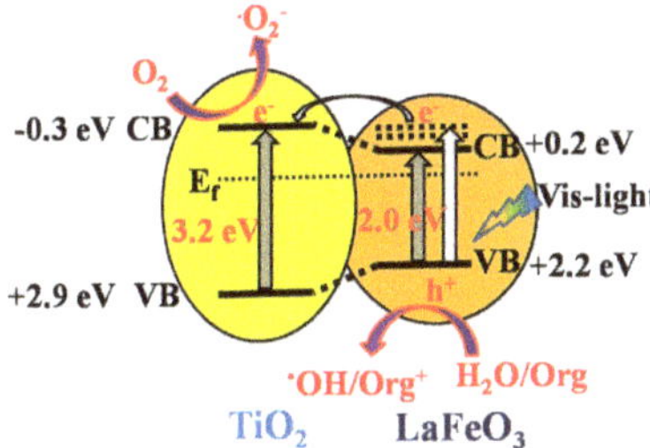

Figure 7. Scheme for energy band gaps and the mechanism for photogenerated charge separation and transfer in the fabricated T/P-LFO nanocomposite.

3. Experimental Section

3.1. Chemicals and Reagents

All the reagents were of analytical grade and used as received without further purification. Deionized water was used throughout the experiment.

3.2. Synthesis of Porous $LaFeO_3$

Porous $LaFeO_3$ nanoparticles were synthesized by taking equimolar amounts (0.04 mol) of $La(NO_3)\cdot 6H_2O$ and $Fe(NO_3)_3\cdot 9H_2O$ and dissolved into a mixed solvent of ethylene glycol (EG) and methanol (3:7 vol %) at room temperature. The solution was treated by ultrasonication for 1 h. Then polystyrene (PS) colloidal crystals were soaked in the precursor solution and kept under vigorous magnetic stirring for 12 h. After that, the mixture was dried in oven at 80 °C for 12 h. The dry powder was then calcined in air at 400 °C (temp ramp 1 $°C\cdot min^{-1}$) for 2 h to remove the polystyrene spheres. Finally the product was calcined at 600 °C (5 $°C\cdot min^{-1}$) for 2 h to obtain porous $LaFeO_3$ nanoparticles.

3.3. Fabrication of Nanocomposite Materials

To fabricate different T/P-LFO nanocomposites, for each sample 1 g freshly prepared P-LFO nanoparticles were taken and suspended into a mixed solvent containing 10 mL water, 40 mL anhydrous ethanol, and 2 mL HNO_3 (68%), under vigorous stirring at room temperature. The reaction mixtures were treated by ultrasonication for 10 min. Then the mixtures were kept under vigorous stirring for 30 min. After that, a certain volume (0.3, 0.6, 0.9, 1.2 mL) of the mixed solution of Titanium butoxide $Ti[(OCH_2)_3CH_3]_4$ and anhydrous ethanol with a ratio of (1:9 vol %) was added to the reaction mixtures under vigorous stirring for 2 h. Subsequently, the mixtures were dried at 85 °C in an oven, followed by calcining at 450 °C for 2 h. Different mole ratios of T/P-LFO nanocomposites were obtained and denoted by X T-LF, where X represents the mole ratio percentage of Ti to La.

3.4. Evaluation of Photocatalytic Activity for Pollutant Degradation

Here in this work, liquid-phase phenol and gas-phase acetaldehyde have been chosen as model pollutants because phenol is a typical recalcitrant contaminant without sensitizing as a dye and acetaldehyde is a kind of volatile organic compound which widely exist in industrial production and are harmful to both human health and natural environment. Therefore, both the pollutants were selected to evaluate the visible-light photocatalytic activities of the fabricated T/P-LFO nanocomposites. The liquid-phase photocatalytic experiments were carried out in an open photochemical reactor glass with 100 mL volume under visible irradiation with cutoff 420 nm wavelength filter using a source of 150 W GYZ220 high-pressure Xenon lamp made in China. The distance of the light source from the reactor glass was approximately 10 cm. In a typical experiment, 0.1 g of photocatalyst and 80 mL of 10 mg/L phenol solution were mixed under stirring for 0.5 h, so as to reach the adsorption saturation and then irradiated under visible-light for 1 h. After irradiation for 1 h, the solution was centrifuged and the concentration of phenol was analyzed with a Model Shimadzu UV-2550 Spectrophotometer (Kyoto, Japan) using colorimetric method of 4-aminoantipyrine at the characteristic optical adsorption of 510 nm.

For the photocatalytic degradation of gas-phase acetaldehyde, the experiments were carried out in a 3 mouth cylindrical quartz reactor with 640 mL volume. A desired amount of photocatalyst and a specific concentration of acetaldehyde gas is introduced through these mouths. After carefully sealed, the reactor was irradiated under visible-light using a source of 150 W Xenon lamp with cutoff filter ($\lambda > 420$ nm). In a typical experiment, 0.1 g of photocatalyst was suspended in a quartz reactor containing a mixture of 810 ppm acetaldehyde, 20% O_2 and 80% of N_2. Prior to irradiation, the gases were mixed by continuous flow through the reactor for half an hour to reach the adsorption saturation. The concentration of acetaldehyde in the photocatalysis was detected with a gas chromatograph (GC-2014, Shimadzu, Kyoto, Japan) equipped with a flame ionization detector. After that, the sample was irradiated by visible-light for 1 h and the concentration of the acetaldehyde gas was re-measured.

3.5. Measurement of the Produced Hydroxyl Radical (•OH) Amount

To measure the amount of hydroxyl radicals, in a typical experiment, 50 mg of photocatalyst and 20 mL of 5 $mg \cdot L^{-1}$ aqueous solution of coumarin were mixed in a 50 mL of quartz glass reactor. The reactor was irradiated under visible-light using a source of 150 W high-pressure Xenon lamp with cutoff filter ($\lambda > 420$ nm) under continuous magnetic stirring for 1 h. The distance of the source from the reactor was about 10 cm. After irradiation for 1 h, a certain amount of solution was taken in a Pyrex glass cell for the fluorescence measurement of 7-hydroxycoumarin under excitation wavelength of 350 nm.

3.6. Characterization

The materials were characterized by using various techniques. The crystal structures of the samples were determined with the help of XRD (Rigaku D/MAX-RA diffractometer, Kyoto, Japan), operated at an accelerating voltage of 30 kV, using Cu Kα radiation ($\alpha = 0.15418$ nm). During measurement, the emission current of 20 mA was employed. The UV-vis diffuse reflectance spectra of the samples were obtained with the help of Shimadzu UV-2550 spectrophotometer (Kyoto, Japan), using $BaSO_4$ as a reference. Transmission electron microscopy (TEM) micrographs of the samples were taken by a JEOL, JEM-2100 (Tokyo, Japan), electron microscope operated at an acceleration voltage of 300 kV. Scanning electron microscopy (SEM) images were taken using a Hitachi S-4800 instrument (Tokyo, Japan), operating at acceleration voltage of 15 kV. The chemical compositions and elemental states of the samples were tested with the help of Kratos-Axis Ultra DLD X-ray photoelectron spectroscopy (XPS) (Kyoto, Japan), with an Al (mono) X-ray source and the binding energies of the samples were calibrated with respect to the signal for adventitious carbon (binding energy = 284.6 eV). The N_2 adsorption–desorption isotherm of various samples were carried out by Micromeritics Tristar II 3020 system (Atlanta, GA, USA) at the temperature of liquid nitrogen, while keeping the system out-gassed for 10 h at 150 °C prior to measurements. The photo luminescence (PL) spectra of the samples were measured with a PE LS 55 spectrofluorophotometer (Waltham, MA, USA) at excitation wavelength of 325 nm.

The atmosphere-controlled surface photovoltage spectroscopy (SPS) spectra of the samples were detected with a home-built apparatus, equipped with a lock-in amplifier (SR830, Sunnyvale, CA, USA) synchronized with a light chopper (SR540, Sunnyvale, CA, USA). The powder samples were sandwiched between two indium tin oxide (ITO) glass electrodes, which were fixed in an atmosphere-controlled system with a quartz window. The monochromatic light was obtained by passing light from a source of 500W Xenon lamp (CHF XQ500W, Global xenon lamp power) through a double prism monochromator (SBP3000, Beijing, China).

4. Conclusions

We have developed T/P-LFO nanocomposites, which exhibit superior visible-light photocatalytic activity for acetaldehyde and phenol degradation. The enhanced photocatalytic activities of T/P-LFO nanocomposites can be attributed to the effective electron-hole pair separation by transferring electrons from P-LFO to TiO_2. This research demonstrates that T/P-LFO nanocomposites show promising applications in the field of photocatalysis by utilizing solar energy.

Acknowledgments: We are grateful for financial support from the NSFC project (U1401245, 21501052), the National Key Basic Research Program of China (2014CB660814), the Program for Innovative Research Team in Chinese Universities (IRT1237), the Research Project of Chinese Ministry of Education (213011A), the Specialized Research Fund for the Doctoral Program of Higher Education (20122301110002), the Science Foundation for Excellent Youth of Harbin City of China (2014RFYXJ002). Special thanks to Chinese scholarship council for financial support.

Author Contributions: The experimental design was planned by Liqiang Jing and Yang Qu. The experimental work and data analysis were performed by Muhammad Humayun, Zhijun Li, Liqun Sun, Xuliang Zhang, Fazal Raziq and Amir Zada.

Conflicts of Interest: The authors declare no conflict of interest.

References

1. Li, Z.X.; Shen, Y.; Yang, C.; Lei, Y.C.; Guan, Y.; Lin, Y.; Liu, D.; Nan, C.W. Significant Enhancement in the Visible Light Photocatalytic Properties of $BiFeO_3$-Graphene Nanohybrids. *J. Mater. Chem. A* **2013**, *1*, 823–829. [CrossRef]
2. Xie, M.Z.; Fu, X.D.; Jing, L.Q.; Luan, P.; Feng, Y.J.; Fu, H.G. Long-Lived, Visible-Light-Excited Charge Carriers of $TiO_2/BiVO_4$ Nanocomposites and Their Unexpected Photoactivity for Water Splitting. *Adv. Energy Mater.* **2014**, *4*. [CrossRef]
3. Fujii, T.; Matsusue, I.; Nakatsuka, D.; Nakanishi, M.; Takada, J. Synthesis and anomalous magnetic properties of $LaFeO_3$ nanoparticles by hot soap method. *Mater. Chem. Phys.* **2011**, *129*, 805–809. [CrossRef]
4. Zhu, J.J.; Li, H.L.; Zhong, L.Y.; Xiao, P.; Xu, X.L.; Yang, X.G.; Zhao, Z.; Li, J.L. Perovskite Oxides: Preparation, Characterizations, and Applications in Heterogeneous Catalysis. *ACS Catal.* **2014**, *4*, 2917–2940. [CrossRef]
5. Kanade, K.G.; Baek, J.O.; Kong, K.J.; Kale, B.B.; Lee, S.M.; Moon, S.J.; Lee, C.W.; Yoon, S.H. A new layer perovskites $Pb_2Ga_2Nb_2O_{10}$ and $RbPb_2Nb_2O_7$: An efficient visible light driven photocatalysts to hydrogen generation. *Int. J. Hydrog. Energy* **2008**, *33*, 6904–6912. [CrossRef]
6. Hirohisa, T.; Makoto, M. Advances in designing perovskite catalysts. *Curr. Opin. Solid State Mater. Sci.* **2001**, *5*, 381–387.

7. Abazari, R.; Sanati, S. Perovskite $LaFeO_3$ nanoparticles synthesized by the reverse microemulsion nanoreactors in the presence of aerosol-OT: Morphology, crystal structure, and their optical properties. *Superlattices Microst.* **2013**, *64*, 148–157. [CrossRef]
8. May, K.J.; Fenning, D.P.; Ming, T.; Hong, W.T.; Lee, D.Y.; Stoerzinger, K.A.; Biegalski, M.D.; Kolpak, A.M.; Yang, S.H. Thickness-Dependent Photoelectrochemical Water Splitting on Ultrathin $LaFeO_3$ Films Grown on Nb:$SrTiO_3$. *J. Phys. Chem. Lett.* **2015**, *6*, 977–985. [CrossRef]
9. Jing, L.Q.; Qu, Y.C.; Su, H.J.; Yao, C.H.; Fu, H.G. Synthesis of High-Activity TiO_2-Based Photocatalysts by Compounding a Small Amount of Porous Nanosized $LaFeO_3$ and the Activity-Enhanced Mechanisms. *J. Phys. Chem. C* **2011**, *115*, 12375–12380. [CrossRef]
10. Farhadi, S.; Siadatnasab, F. Perovskite-type $LaFeO_3$ nanoparticles prepared by thermal decomposition of the $La[Fe(CN)_6]\cdot 5H_2O$ complex: A new reusable catalyst for rapid and efficient reduction of aromatic nitro compounds to arylamines with propan-2-ol under microwave irradiation. *J. Mol. Catal. A* **2011**, *339*, 108–116. [CrossRef]
11. Yu, Q.; Meng, X.G.; Wang, T.; Li, P.; Liu, L.Q.; Chang, K.; Liu, G.G.; Ye, J.H. A highly durable p-$LaFeO_3$/n-Fe_2O_3 photocell for effective water splitting under visible light. *Chem. Commun.* **2015**, *51*, 3630–3633. [CrossRef]
12. Qu, Y.; Zhou, W.; Xie, Y.; Jiang, L.; Wang, J.; Tian, G.; Ren, Z.; Tian, C.; Fu, H. A novel phase-mixed $MgTiO_3$–$MgTi_2O_5$ heterogeneous nanorod for high efficiency photocatalytic hydrogen production. *Chem. Commun.* **2013**, *49*, 8510–8512. [CrossRef]
13. Varaprasad, K.; Ramam, K.; Reddy, G.S.M.; Sadiku, R. Development and characterization of nano-multifunctional materials for advanced applications. *RSC Adv.* **2014**, *4*, 60363–60370. [CrossRef]
14. Luan, P.; Xie, M.Z.; Fu, X.; Qu, Y.; Sun, X.; Jing, L. Improved photoactivity of TiO_2–Fe_2O_3 nanocomposites for visible-light water splitting after phosphate bridging and its mechanism. *Phys. Chem. Chem. Phys.* **2015**, *17*, 5043. [CrossRef]
15. Xu, J.J.; Wang, Z.L.; Xu, D.; Meng, F.Z.; Zhang, X.B. 3D ordered macroporous $LaFeO_3$ as efficient electrocatalyst for Li–O_2 batteries with enhanced rate capability and cyclic performance. *Energy Environ. Sci.* **2014**, *7*, 2213–2219. [CrossRef]
16. Zhao, J.; Liu, Y.P.; Li, X.W.; Lu, G.Y.; You, L.; Liang, X.; Liu, F.; Zhang, T.; Du, Y. Highly sensitive humidity sensor based on high surface area mesophorous $LaFeO_3$ prepared by a nanocasting route. *Sens. Actuators B* **2013**, *181*, 802–809. [CrossRef]
17. Wang, P.; Zhai, Y.M.; Wang, D.J.; Dong, S.J. Synthesis of reduced graphene oxide-anatase TiO_2 nanocomposite and its improved photo-induced charge transfer properties. *Nanoscale* **2011**, *3*, 1640–1645. [CrossRef]
18. Su, H.J.; Jing, L.Q.; Shi, K.Y.; Yao, C.H.; Fu, H.G. Synthesis of large surface area $LaFeO_3$ nanoparticles by SBA-16 template method as high active visible photocatalysts. *J. Nanopart. Res.* **2010**, *12*, 967–974. [CrossRef]
19. Parida, K.M.; Reddy, K.H.; Martha, S.; Das, D.P.; Biswal, N. Fabrication of nanocrystalline $LaFeO_3$: An efficient sole gel auto-combustion assisted visible light responsive photocatalyst for water decomposition. *Int. J. Hydrog. Energy* **2010**, *35*, 12161–12168. [CrossRef]
20. Li, Z.L.; Wnetrzak, R.; Kwapinski, W.; Leahy, J.J. Synthesis and characterization of sulfated TiO_2 nanorods and ZrO_2/TiO_2 nanocomposites for the esterification of bio-based organic acid. *ACS Appl. Mater. Interfaces* **2012**, *4*, 4499–4505. [CrossRef]
21. Song, P.; Zhang, H.H.; Han, D.; Li, J.; Yang, Z.X.; Wang, Q. Preparation of biomorphic porous $LaFeO_3$ by sorghum straw biotemplate method and its acetone sensing properties. *Sens. Actuators B* **2014**, *196*, 140–146. [CrossRef]
22. Gao, K.; Li, S.D. Multi-modal TiO_2–$LaFeO_3$ composite films with high photocatalytic activity and hydrophilicity. *Appl. Surf. Sci.* **2012**, *258*, 6460–6464. [CrossRef]
23. Humayun, M.; Zada, A.; Li, Z.J.; Xie, M.Z.; Zhang, X.; Qu, Y.; Raziq, F.; Jing, L.Q. Enhanced visible-light activities of porous $BiFeO_3$ by coupling with nanocrystalline TiO_2 and mechanism. *Appl. Catal. B* **2016**, *180*, 219–226. [CrossRef]

Article

Mechanochemical Synthesis of TiO_2 Nanocomposites as Photocatalysts for Benzyl Alcohol Photo-Oxidation

Weiyi Ouyang [1], Ewelina Kuna [2], Alfonso Yepez [1], Alina M. Balu [1], Antonio A. Romero [1], Juan Carlos Colmenares [2] and Rafael Luque [1,*]

1 Department of Organic Chemistry, University of Cordoba, Edificio Marie Curie(C-3), Ctra Nnal IV-A, Km 396, Cordoba E14014, Spain; qo2ououw@uco.es (W.O.); z22yegaa@uco.es (A.Y.); qo2balua@uco.es (A.M.B.); qo1rorea@uco.es (A.A.R.)

2 Institute of Physical Chemistry Polish Academy of Sciences (PAS), Kasprzaka 44/52, Warsaw 01-224, Poland; ekuna@ichf.edu.pl (E.K.); jcarloscolmenares@ichf.edu.pl (J.C.C.)

* Correspondence: q62alsor@uco.es; Tel.: +34-957211050

Academic Editors: Hermenegildo García and Sergio Navalón
Received: 3 March 2016; Accepted: 7 May 2016; Published: 18 May 2016

Abstract: TiO_2 (anatase phase) has excellent photocatalytic performance and different methods have been reported to overcome its main limitation of high band gap energy. In this work, TiO_2-magnetically-separable nanocomposites (MAGSNC) photocatalysts with different TiO_2 loading were synthesized using a simple one-pot mechanochemical method. Photocatalysts were characterized by a number of techniques and their photocatalytic activity was tested in the selective oxidation of benzyl alcohol to benzaldehyde. Extension of light absorption into the visible region was achieved upon titania incorporation. Results indicated that the photocatalytic activity increased with TiO_2 loading on the catalysts, with moderate conversion (20%) at high benzaldehyde selectivity (84%) achieved for 5% TiO_2-MAGSNC. These findings pointed out a potential strategy for the valorization of lignocellulosic-based biomass under visible light irradiation using designer photocatalytic nanomaterials.

Keywords: TiO_2; magnetically separable photocatalysts; selective photo-oxidation; mechanochemical synthesis; ball mill

1. Introduction

Photocatalysis has been considered as one of the most environmentally friendly and promising technologies owing to advantages such as being clean, efficient, cost-effective, and energy-saving [1–3] Typical applications of photocatalysis are conversion of CO_2 to fuels and chemicals [4–8], self-cleaning surfaces [9,10], disinfection of water [11,12], oxidation of organic compounds [13–15], and production of hydrogen from water splitting [16–19]. In this regard, different types of heterogeneous photocatalysts have been extensively reported, including metal oxide nanoparticles, composite nanomaterials, metal-organic frameworks, plasmonic photocatalysts, and polymeric graphitic carbon nitride [3,4].

Among these different types of photocatalysts, TiO_2 has been extensively investigated and is one of the most widely used in the aforementioned applications due to its excellent photocatalytic activity, high thermal and chemical stability, low cost, and non-toxicity [20,21]. However, in spite of its advantages, the main drawback of TiO_2 in photocatalysis relates to the large band gap (3.2 eV) for its anatase crystalline phase which restricts its utilization to ultraviolet (UV) irradiation ($\lambda < 387$ nm), with UV irradiation comprising less than 5% of the solar energy. Therefore, it is very important to extend the photocatalytic activity of TiO_2 nanocatalysts under visible light to profit from abundant solar energy. Various approaches have been developed to improve the photoactivity of TiO_2 by lowering the band-gap energy and delaying the recombination of the excited electron-hole pairs, *i.e.*, cationic [22,23] and anionic [20,24,25] doping, dye photosensitization, deposition of noble metals. Photocatalysts doped with noble metals can improve their photoactivities, but with limitations for large scale applications. Importantly, the design of photocatalysts featuring magnetic separation has not been considered to a large extent despite the obvious advantages of separation and recycling for magnetically-separable heterogeneous photocatalysts [26]. Conventional methods for heterogeneous catalyst recovery, such as filtration, centrifugation, *etc.*, are either time consuming or costly, while the enhanced magnetically-separable properties of the heterogeneous catalyst can exceed these limitations. In recent years, photocatalysts with TiO_2 coated on magnetic particles have been reported by many researchers [27–29], which showed enhanced photocatalytic activities and feasible separation by applying external magnetic field. Ojeda *et al.* reported a maghemite/silica nanocomposites, which were also magnetically separable [30], followed by a report on the incorporation of TiO_2 on maghemite/silica nanocomposites under ultrasounds which exhibited excellent photocatalytic performance in the selective oxidation of benzyl alcohol [31].

The selective oxidation of alcohols to the corresponding carbonyl compounds accounts for one of the most significant transformations in organic chemistry. Particularly, the conversion of benzyl alcohol (BA) to benzaldehyde (BHA) has attracted extensive attention, since benzaldehyde is widely applied in food, pharmaceutical, and perfumery industries and as building block in other chemical industries. Recently, the photocatalytic oxidation of benzyl alcohol to benzaldehyde has been reported using different catalysts and chlorine-free benzaldehyde with high selectivity, with respect to the traditional syntheses-either by benzyl chloride hydrolysis or via toluene oxidation [15,26,32].

In continuation with research efforts from the group related to the design of advanced nanomaterials for (photo)catalytic processes, we aimed to synthesize an advanced magnetically-separable nanophotocatalyst (TiO_2-MAGSNC) using a simple one-pot mechanochemical method under ball mill. A widely-reported porous support (SBA-15) was utilized as support, together with an iron precursor and propionic acid to obtain a magnetic phase able to provide magnetically-separable features to the catalyst. A high-energy ball milling process was applied in this work which could provide small nanoparticle sizes as well as a highly homogeneous crystalline structure and morphology. TiO_2-MAGSNC catalysts were found to be photoactive with a high selectivity in the selective oxidation of benzyl alcohol to benzaldehyde.

2. Experimental

2.1. Synthesis of TiO_2/MAGSNC Photocatalysts

SBA-15 silica was prepared using the procedure reported by Bonardet *et al.* [33] Different amounts of titanium precursor were used to obtain various contents of TiO_2 (0.5, 1.0, 2.0, 5.0 wt %) on the catalysts. Titanium incorporation was subsequently achieved by a simple mechanochemical method in a planetary ball mill under previous optimized conditions [34]. In detail, Pluronic P123 surfactant (Sigma-Aldrich Inc., St. Louis, MO, USA) (8.0 g) was dissolved in deionized water (260 mL) and HCl (Panreac Química S.L.U., Barcelona, Catalonia, Spain) (12 M, 40 mL) under vigorous stirring, at 40 °C for 2 h. Upon complete dissolution, 7 g of tetraethyl orthosilicate (TEOS) (Sigma-Aldrich Inc., St. Louis, MO, USA) were added dropwise to the above solution. The mixture was stirred at 40 °C for 24 h, followed by hydrothermal treatment at 100 °C for 48 h in an oven. The white solid was separated from the solution by filtration and dried at 60 °C. The template was removed by calcination at 600 °C for 8 h. Different amounts (13, 59, 188 and 661 μL) of titanium isopropoxide (Sigma-Aldrich Inc., St. Louis, MO, USA), 1.34 g $Fe(NO_3)_3 \cdot 9H_2O$ (Merck, Darmstadt, Hesse, Germany), 0.5 g SBA-15 and 0.25 mL propionic acid (Panreac Química S.L.U., Barcelona, Catalonia, Spain) were added to a 125 mL reaction chamber with eighteen 10 mm stainless steel balls and then ground in a Retsch PM-100 planetary ball mill (350 rpm, 10 min) (Retsch GmbH, Haan, North Rhine-Westphalia, Germany). Materials calcination was performed at 400 °C (heating rate 3 °C/min) for 5 h in a furnace under an oxygen deficient atmosphere (static air). MAGSNC sample was synthesized under same conditions without adding titanium isopropoxide. All chemicals were used as received.

2.2. Characterization of the TiO_2-MAGSNC Photocatalysts

The crystal phase structures of TiO_2-MAGSNC samples were examined by powder X-ray diffraction (XRD) measurements performed in a Bruker D8 DISCOVER A25 diffractometer (Bruker Corporation, Billerica, MA, USA) equipped with a vertical goniometer under theta-theta geometry using Ni filtered Cu Kα (λ = 1.5418 Å) radiation and operated at 40 KeV and 40 mA. Wide angle scanning patterns were collected from 10° to 80° with a step size of 0.01° and counting time of 500 s per step.

Textural properties of the samples were determined by N_2 physisorption using a Micromeritics ASAP 2020 automated system (Micromeritics Instrument Corporation, Norcross, GA, USA) with the Brunauer-Emmet-Teller (BET) and the Barret-Joyner-Halenda (BJH) methods. Prior to adsorption measurements, samples were degassed under vacuum (0.1 Pa) for 4 h at 300 °C.

A UV/VIS/NIR spectrophotometer Jasco V-570 (JASCO international Co., Ltd., Hachioji, Tokyo, Japan) equipped with an integrating sphere was used to record Ultraviolet-Visible (UV-VIS) diffuse reflectance spectra. The baseline was obtained with SpectralonTM (poly(tetrafluoroethylene) as a reference material. The Kubelka-Munk method was utilized (from diffuse reflectance spectra) to determine the band gap function. Function $f(R)$ was calculated from the following equation:

$$f(R) = \frac{(1-R)^2}{2R} \quad (1)$$

while E_g was calculated from $(f(R)h\nu)^{1/2}$ *versus* hν plots.

X-ray photoelectron spectroscopy (XPS) measurements were carried out with a VG Scientific photoelectron spectrometer ESCALAB-210 (Thermo Scientific, Waltham, MA, USA) with Al Kα radiation (1486.6 eV) from an X-ray source, operating at 15 kV and 20 mA. Survey spectra in the energy range from 0 to 1350 eV with 0.4 eV step were recorded for all the samples. High resolution spectra were recorded with 0.1 eV step, 100 ms dwell time and 25 eV pass energy. A ninety degree take-off angle was employed in all measurements. Curve fitting was carried out using the CasaXPS software (Casa Software Ltd., Cheshire, England, UK), which each component of the complex envelope is described as a Gaussian–Lorentzian sum function; a constant 0.3 (±0.05) G/L ratio was used. The background was fitted using a nonlinear Shirley model. Measured transmission function and Scofield sensitivity factors have been employed for quantification purposes. An aromatic carbon C 1s peak at 284.5 eV was used as the reference of binding energy.

Scanning electron microscopy images were recorded with a JEOL JSM-6300 scanning microscope (JEOL Ltd., Akishima, Tokyo, Japan) equipped with Energy-dispersive X-ray spectroscopy (EDX) at 20 kV. An Au/Pd coating was employed to analyze samples on a high-resolution sputtering SC7640 instrument (Quorum Technologies Ltd., Lewes, England, UK) (up to 7 nm thickness) at a sputtering rate of 1.5 kV per minute.

FEI Tecnai G2 (FEI Tecnai, Hillsboro, OR, USA) fitted with a Charge-coupled Device (CCD) camera for ease and speed of use was applied to record the transmission electron microscopy (TEM) images of the synthesized TiO_2-MAGSNC samples at the Research Support Service Center (SCAI) from Universidad de Cordoba. The resolution of the equipment is around 0.4 nm. Prior to the recording, samples were prepared by suspension in ethanol, assisted by sonication and followed by deposition on a copper grid.

The magnetic susceptibility was measured at low frequency (470 Hz) using a Bartington MS-2 (Bartington Instruments Ltd., Witney, England, UK), at room temperature.

2.3. Photocatalytic Experiments

A Pyrex cylindrical double-wall immersion well reactor equipped with medium pressure 125 W mercury lamp (λ = 365 nm), which was supplied by Photochemical Reactors Ltd. UK (Model RQ 3010), (Reading, UK) was used in all the catalytic reactions (Figure 1). The distance between the light source and reaction media was *ca.* (*ca.*: abbreviation of circa) 10 nm and irradiance of the light source reached 1845.6 W/m^2. Magnetic stirring with a speed of 1100 rpm was utilized in the batch reactor to obtain a homogenous suspension of the TiO_2-MAGSNC photocatalysts. The reaction temperature was established at 30 °C. 1.5 mM benzyl alcohol (Sigma-Aldrich Inc., St. Louis, MO, USA) was prepared in acetonitrile (Sigma-Aldrich Inc., St. Louis, MO, USA) medium. Experiments were performed from 150 mL of the mother solution and 1 g/L of catalyst concentration for 4 h under UV light and air bubbling conditions (25 mL/min). In order to equilibrate the adsorption-desorption over the photocatalyst surface, the reaction solution was left in the dark for 30 min before each reaction. Samples were periodically withdrawn (*ca.* 1 mL) from the photoreactor at different times and filtered off (0.20 μm, 25 mm, nylon filters). The concentration of model compound was determined by a high performance liquid chromatography (HPLC, Waters Model 590 pump) (Waters Limited, Hertfordshire, UK) equipped with a dual absorbance detector (Waters 2487) and the SunFire™ C18 (3.5 μm, 150 mm length, 4.6 mm inner diameter) column provided by Waters. The mobile phase was Milli-Q water/acetonitrile/methanol in the volumetric ratio of 77.5:20:2.5 with 0.1% of H_3PO_4 (Sigma-Aldrich Inc., St. Louis, MO, USA). We used isocratic elution at a flow rate of 1 mL/min. The injection volume was 10 μL. TiO_2 P25 (approx. 80% anatase and 20% rutile) is a commercial catalyst purchased from Evonik Industries (Evonik Industries AG, Essen, Germany) and used as comparison here.

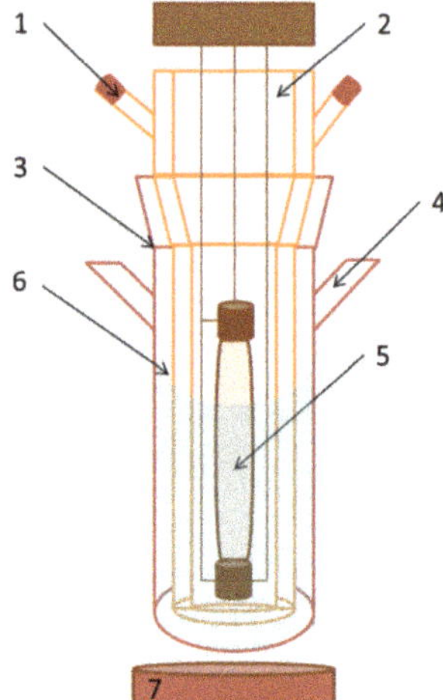

Figure 1. Reaction system: (1) lamp cooling system; (2) double-walled immersion well reactor; (3) photoreactor; (4) port for taking samples; (5) 125 W ultraviolet (UV) lamp; (6) mother solution; and (7) magnetic stirrer.

3. Results and Discussion

XRD analysis was performed to investigate the crystal phase of the synthesized TiO_2-MAGSNC nanocomposites. The XRD pattern of a representative sample (5% TiO_2-MAGSNC) is shown in Figure 2 The mean observed peaks ($2\theta = 35.6°$) could be assigned to the presence of a magnetic phase (in principle γ-Fe_2O_3, although the presence of a magnetite phase cannot be completely ruled out) while titania peaks were not obvious due to the low titanium loading on the supports; hence, particle size could not be worked out from these data. By applying the Scherrer equation, iron oxide nanoparticle sizes can be calculated to be *ca.* 14 nm. Results from XRD pattern also suggested that our simple mechanochemical protocol can successfully lead to the formation of magnetically-separable nanocomposites, as further supported with subsequent characterization techniques.

N_2 absorption-desorption isotherms were used to evaluate the textural properties of the TiO_2-MAGSNC samples with different content of TiO_2. The isotherms (Figure 3) matched the characteristic type IV isotherm profile indicating these samples are essentially mesoporous in nature. In comparison to commercial titanium oxide (59 $m^2 \cdot g^{-1}$) our materials possess significantly higher surface area (generally 400–500 $m^2 \cdot g^{-1}$), without any significant changes in terms of textural properties with respect to those of the parent MAGSNC, probably due to the low titania loading. These could also be observed in TEM images. Pore volumes in the 0.40–0.45 mLg^{-1} range and diameters typical of the parent SBA-15 material (*ca.* 6 nm) were also obtained.

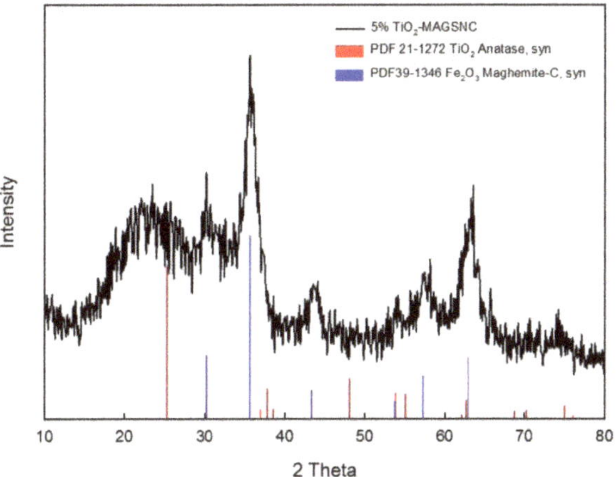

Figure 2. X-ray diffraction (XRD) pattern of 5% TiO_2-magnetically-separable photocatalysts (MAGSNC) photocatalysts. (PDF 21-1272 and PDF 39-1346 are the card numbers for the crystalline structures in the data base, while Anatase, syn and Maghemite-C, syn are the corresponding structure names.)

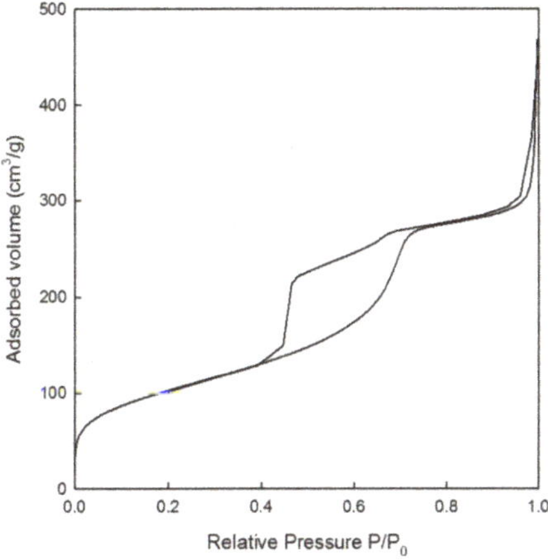

Figure 3. N_2 absorption-desorption isotherm of 5% TiO_2-MAGSNC photocatalysts. P: partial vapor pressure of adsorbate gas in equilibrium with the surface at 77.4 K; P_0: saturated pressure of adsorbate gas.

Diffuse reflectance (DR) UV-VIS spectroscopy was used to record the optical properties of the samples. UV-VIS adsorption spectra of TiO_2-MAGSNC samples are shown in Figure 4, which showed extensions of absorption band into the visible region for all catalysts. Significant enhancement of light absorption of all samples was achieved at a wavelength of around 700 nm, when comparing to those of pure commercial TiO_2 (P25, 386 nm). The extension of light absorption of the synthesized catalysts into the visible range was probably resulting from the presence of the photocatalytic composite, iron oxide phase, on the MAGSNC supports. As a result of the extension of light absorption into the visible light range, better utilization of the abundant solar energy might be possible.

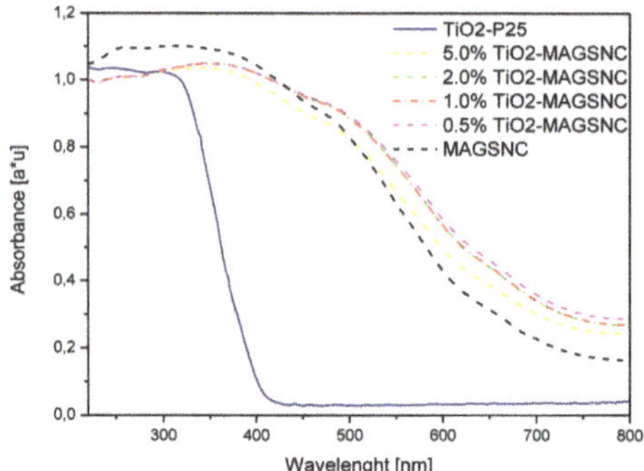

Figure 4. Diffuse reflectance (DR) Ultraviolet-Visible (UV-VIS) absorption spectra of different TiO_2-MAGSNC photocatalysts. P25: pure commercial TiO_2 from Evonik Industries.

The band gaps of synthesized TiO_2-MAGSNC were calculated, based on the Kubelka-Munk function (Table 1), to be in the 1.62 to 1.67 eV range. These extraordinary low values are derived from the iron oxide phase formed during ball mill in the synthetic stage as a result of the mechanochemical process [30,34], which only slightly decrease upon titanium incorporation. With Fe^{3+} radius (0.64 Å) close to that of Ti^{4+} (0.68 Å), the incorporation of Fe^{3+} into the TiO_2 crystal lattice during synthesis may also take place [35]. The proposed one-pot synthesis procedure might facilitate the incorporation of Fe^{3+} and formation of heterojunctions between TiO_2 and iron oxide phases during the transformation of titanium precursor to TiO_2 which might favor the charge separation in the catalysts and further improve the photocatalytic activity.

Table 1. Optical properties of synthesized TiO_2-MAGSNC photocatalysts. P25: pure commercial TiO_2 from Evonik Industries.

Materials	Band Gap [eV]	Absorption Threshold [nm]
TiO_2-P25	3.21	386
MAGSNC	1.75	705
0.5% TiO_2-MAGSNC	1.62	765
1.0% TiO_2-MAGSNC	1.63	761
2.0% TiO_2-MAGSNC	1.65	751
5.0% TiO_2-MAGSNC	1.67	740

In order to analyze the chemical states of the prepared samples, XPS spectra were also recorded. Figure 5a depicts binding energies (BEs) of *ca.* 463.3 and 457.5 eV for Ti 2p3/2 and Ti 2p1/2, respectively, characteristic of the Ti^{4+} cation with a 5.8 eV spin orbit splitting. The fitting peak with higher binding energy arises from the Ti^{4+} species in a Ti–O–Fe structure. Electrons can be induced by transfer from Ti^{4+} to Fe^{3+} in the Ti–O–Fe bond due to the electronegativity difference between Ti^{4+} (1.54) and Fe^{3+} (1.83), which makes Ti^{4+} species potentially less electron-rich (and Fe^{3+} more electron-rich), resulting in the increase of BE for Ti^{4+} species and decrease of BE for Fe^{3+} [36]. Peaks at a binding energy of 723.8 (Fe 2p1/2) and 710.2 eV (Fe 2p3/2) also correlated well to typical signals of Fe^{3+} from Fe 2p in Figure 5b, which confirmed the presence of such species in the nanocomposites, in good agreement with XRD results. Despite the stability of the hematite phase (as most thermodynamically stable at temperature over 300 °C), the magnetic phase was still well preserved after calcination at 400 °C. Most importantly, the absence of any Fe^{2+} species on the external surface in all catalysts can be confirmed from XPS spectra (Figure 5).

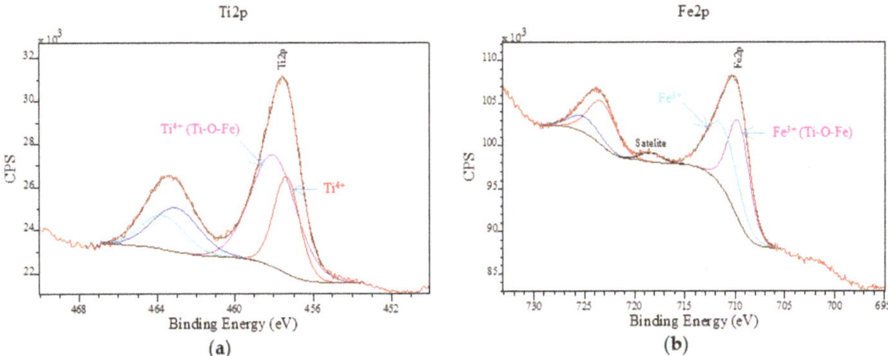

Figure 5. X-ray photoelectron spectroscopy (XPS) spectra of 5% TiO_2-MAGSNC photocatalysts: (**a**) Ti 2p; and (**b**) Fe 2p.

Both scanning electron microscopy (SEM) and TEM images of the catalysts were in good agreement with the textural properties and characterization results of the mesoporous nanocomposites (Figures 6 and 7). Element mapping illustrated for 2% TiO_2-MAGSNC pointed out that both Ti^{4+} and Fe^{3+} were homogeneously distributed on the supports, in line with analogous observations for the other catalysts. Particularly, the fully preserved SBA-15 structure could be visualized in TEM micrographs of the final photocatalytic nanomaterials, with small nanoparticles (*ca.* average nanoparticle size 10 nm), in good agreement with XRD results. Titania nanoparticles could not be distinguished from TEM images, in line with XRD data, which may again relate to a very high dispersion of TiO_2 in the nanocomposites at such low loadings. Results of EDX analysis have been summarized on Table 2, showing a good agreement in terms of Ti content on the catalysts with respect to the theoretical Ti content selected. These findings confirm the excellent incorporation of Ti provided by the proposed mechanochemical approach.

Table 2. Ti and Fe content on TiO_2-MAGSNC photocatalysts (obtained from energy-dispersive X-ray spectroscopy (EDX) analysis).

Sample ID	Ti (wt %)	Fe (wt %)
0.5% TiO_2-MAGSNC	0.2	19.2
1.0% TiO_2-MAGSNC	1.0	24.4
2.0% TiO_2-MAGSNC	1.7	16.2
5.0% TiO_2-MAGSNC	4.7	9.6

The magnetic susceptibility of the TiO_2-MAGSNC photocatalysts were summarized in Table 3, showing that the obtained catalysts all possessed relatively strong ferromagnetism and could be easily separated from the reaction mixture using a simple magnet.

Table 3. The magnetic susceptibility of the TiO_2-MAGSNC photocatalysts.

Sample ID	Magnetic Susceptibility ($\times 10^{-6}$ $m^3 \cdot kg^{-1}$)
0.5% TiO_2-MAGSNC	116.7
1.0% TiO_2-MAGSNC	179.1
2.0% TiO_2-MAGSNC	117.7
5.0% TiO_2-MAGSNC	130.0

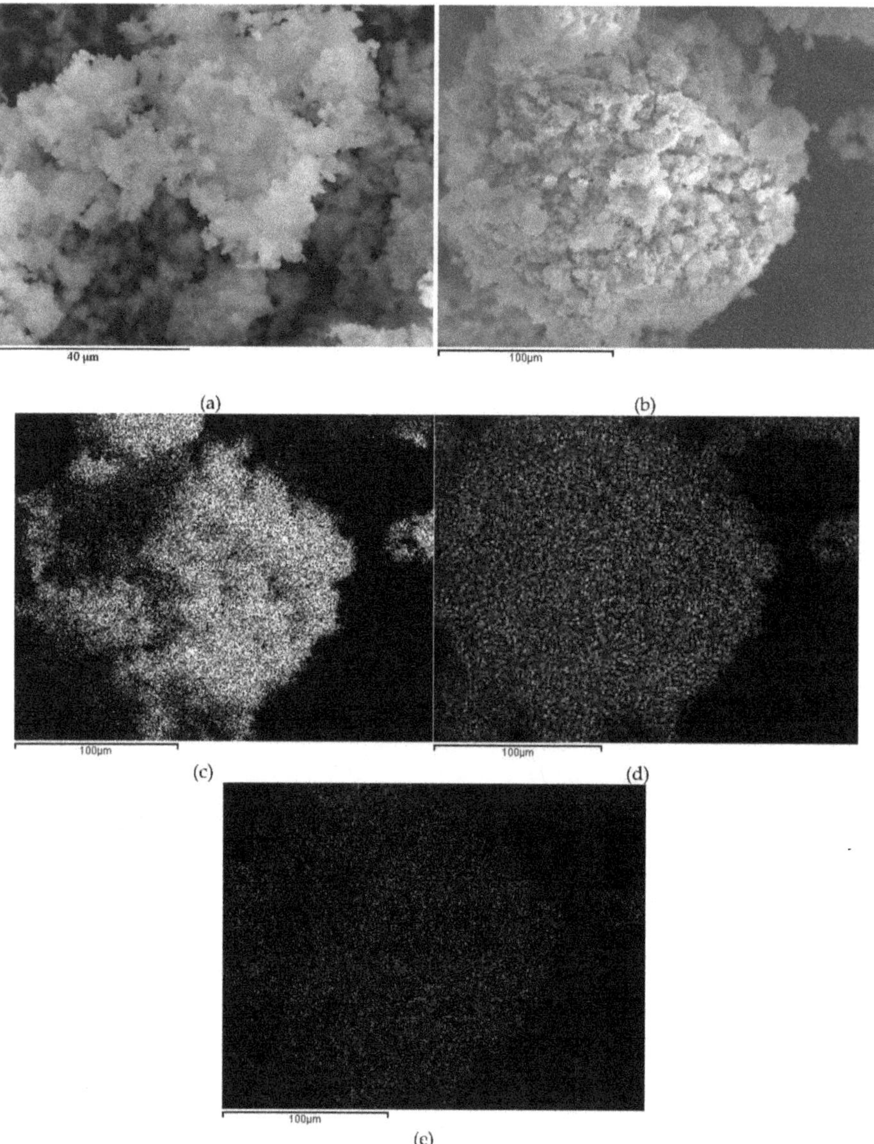

Figure 6. Scanning electron microscopy (SEM) images: (**a**) 5% TiO_2-MAGSNC photocatalysts; (**b**) 2% TiO_2-MAGSNC nanocomposites; and elements mapping of 2% TiO_2-MAGSNC photocatalysts: (**c**) Si; (**d**) Fe; (**e**) Ti.

Figure 7. Transmission electron microscopy (TEM) images of 5% TiO_2-MAGSNC photocatalysts.

After characterization, the effectiveness of TiO_2-MAGSNC photocatalysts with different TiO_2 content was subsequently studied in the photo-oxidation of benzyl alcohol. Photocatalytic activity experiment results have been summarized on Table 4. With illumination time of 4 h, the reaction using TiO_2-MAGSNC photocatalysts with low TiO_2 loading ($\leqslant$1.0 wt %) provided negligible (<5%) photoconversion of benzyl alcohol to benzaldehyde, with conversion only increasing with TiO_2 loading. The magnetically-separable support (MAGSNC) or SBA-15 itself did not provide any photoactivity under otherwise identical reaction conditions. Bare iron oxides can promote the recombination of photogenerated electron-hole pairs, resulting in inactive materials. Interestingly, a titania loading as low as 5% onto MAGSNC containing the iron oxide phase could significantly decrease the band gap of the TiO_2 as well as improve the photoconversion of benzyl alcohol (up to 20% in this work) with a remarkable 84% selectivity to the target product. No over-oxidation products, such as benzoic acid and/or CO_2 from mineralization, were observed in the photo-oxidation of benzyl alcohol photocatalyzed by TiO_2-MAGSNC. Under the same photocatalytic conditions, the photoconversion of P25 Evonik was obviously quantitative but with an extremely low selectivity to benzaldehyde (32%, over 65% to mineralization), almost comparable in terms of product yield. The enhancement of the photocatalytic properties of the TiO_2-MAGSNC catalysts, especially in terms of selectivity, makes very attractive this type of magnetically separable nanocomposite containing low titania content, as compared to pure P25 commercial photocatalysts.

Table 4. Photocatalytic oxidation of benzyl alcohol to benzaldehyde [1].

Catalyst	Conversion [%]	Selectivity BHA [2] [%]	Yield BHA [3] [%]
Blank (no catalyst)	-	-	-
SBA-15	-	-	-
MAGSNC	-	-	-
0.5% TiO_2-MAGSNC	<5	94	-
1.0% TiO_2-MAGSNC	<5	80	-
2.0% TiO_2-MAGSNC	<10	73	-
5.0% TiO_2-MAGSNC	20	84	17
P25 Evonik	>95	32	30

[1] Reaction conditions: C_o benzyl alcohol = 1.5 mM, 125 W lamp, loading: 1 g/L. (solvent: acetonitrile, air flow: 25 mL/min, temperature: 30 °C, reaction time: 4 h). [2] BHA: benzaldehyde. [3] The selectivity of a reaction was estimated as the ratio of the required product to the undesirable product formed during reaction. Yields were calculated as the ratio of the desired product formed to the total stoichiometric amount. Amount of substance (in mol) were determined using high performance liquid chromatography (HPLC) analysis.

4. Conclusions

Magnetically-separable catalysts with different content of TiO_2 were synthesized in a one-pot mechanochemical approach. The synthesized TiO_2-MAGSNC photocatalysts showed great improvement in light absorption into the visible light range (around 700 nm), with an interesting performance in the photocatalytic conversion of benzyl alcohol to benzaldehyde, particularly at higher loadings (5% Ti). The proposed systems will pave the way to further investigations currently ongoing in our group to the design of photoactive nanomaterials for selective oxidations, which will be reported in due course.

Acknowledgments: Rafael Luque gratefully acknowledges Consejeria de Ciencia e Innovacion, Junta de Andalucia for funding project P10-FQM-6711. Funding from Marie Curie Actions under Innovative Training Networks Project Photo4Future (H2020-MSCA-ITN-2014-641861), especially for funding Weiyi Ouyang Ph.D. studies. Juan Carlos Colmenares, Alina M. Balu and Rafael Luque gratefully acknowledge support from COST Action FP1306 for networking and possibilities for meetings and future students exchange. Juan Carlos Colmenares would like to thank for the support from the National Science Centre (Poland) within the project Sonata Bis Nr. 2015/18/E/ST5/00306.

Author Contributions: Weiyi Ouyang was responsible for the synthesis and characterization of the catalysts, also writing and revising the manuscript; Ewelina Kuna was in charge of the photocatalytic activity test; Alfonso Yepez offered his help in the material synthesis and characterization; Alina M. Balu and Antonio A. Romero supervised the work and lead the discussion while Juan Carlos Colmenares supervised the work in Poland and offered information on the photocatalytic experiments; Rafa Luque provided to project concept and was in charge of completing and revising the manuscript from submission to acceptance.

Conflicts of Interest: The authors declare no conflict of interest.

Abbreviations

The following abbreviations are used in this manuscript:

BA	Benzyl alcohol
BE	Binding energy
BHA	Benzaldehyde
ca.	Circa
EDX	Energy-dispersive X-ray spectroscopy
MAGSNC	Magnetically separable nanocomposites
P	Partial vapor pressure of adsorbate gas in equilibrium with the surface at 77.4 K
P_0	Saturated pressure of adsorbate gas
P25	Pure commercial TiO_2 from Evonik Industries
SEM	Scanning electron microscopy
TEM	Transmission electron microscopy
UV- Vis	Ultraviolet- Visible
XPS	X-ray photoelectron spectroscopy
XRD	Powder X-ray diffraction

References

1. Herrmann, J. Heterogeneous photocatalysis: Fundamentals and applications to the removal of various types of aqueous pollutants. *Catal. Today* **1999**, *53*, 115–129. [CrossRef]
2. Colmenares, J.C.; Luque, R. Heterogeneous photocatalytic nanomaterials: Prospects and challenges in selective transformations of biomass-derived compounds. *Chem. Soc. Rev.* **2014**, *43*, 765–778. [CrossRef] [PubMed]
3. Lang, X.; Chen, X.; Zhao, J. Heterogeneous visible light photocatalysis for selective organic transformations. *Chem. Soc. Rev.* **2014**, *43*, 473–486. [CrossRef] [PubMed]
4. Chen, D.; Zhang, X.; Lee, A.F. Synthetic strategies to nanostructured photocatalysts for CO_2 reduction to solar fuels and chemicals. *J. Mater. Chem. A* **2015**, *3*, 14487–14516. [CrossRef]
5. Adachi, K.; Ohta, K.; Mizuno, T. Photocatalytic reduction of carbon dioxide to hydrocarbon using copper-loaded titanium dioxide. *Sol. Energy* **1994**, *53*, 187–190. [CrossRef]
6. Fu, Y.; Sun, D.; Chen, Y.; Huang, R.; Ding, Z.; Fu, X.; Li, Z. An amine-functionalized titanium metal-organic framework photocatalyst with visible-light-induced activity for CO_2 reduction. *Angew. Chem. Int. Ed.* **2012**, *51*, 3364–3367. [CrossRef] [PubMed]
7. Roy, S.C.; Varghese, O.K.; Paulose, M.; Grimes, C.A. Toward Solar Fuels: Photocatalytic Conversion of Carbon Dioxide to Hydrocarbons. *ACS Nano* **2010**, *4*, 1259–1278. [CrossRef] [PubMed]

8. Yu, J.; Low, J.; Xiao, W.; Zhou, P.; Jaroniec, M. Enhanced Photocatalytic CO_2-Reduction Activity of Anatase TiO_2 by Coexposed {001} and {101} Facets. *J. Am. Chem. Soc.* **2014**, *136*, 8839–8842. [CrossRef] [PubMed]
9. Parkin, I.P.; Palgrave, R.G. Self-cleaning coatings. *J. Mater. Chem.* **2005**, *15*, 1689–1695. [CrossRef]
10. Blossey, R. Self-cleaning surfaces-virtual realities. *Nat. Mater.* **2003**, *2*, 301–306. [CrossRef] [PubMed]
11. Sakthivel, S.; Neppolian, B.; Shankar, M.V.; Arabindoo, B.; Palanichamy, M.; Murugesan, V. Solar photocatalytic degradation of azo dye: Comparison of photocatalytic efficiency of ZnO and TiO_2. *Sol. Energy Mater. Sol. Cells* **2003**, *77*, 65–82. [CrossRef]
12. Hoffmann, M.R.; Martin, S.T.; Choi, W.; Bahnemannt, D.W. Environmental Applications of Semiconductor Photocatalysis. *Chem. Rev.* **1995**, *95*, 69–96. [CrossRef]
13. Peral, J.; Ollis, D.F. Heterogeneous photocatalytic oxidation of gas-phase organics for air purification: Acetone, 1-butanol, butyraldehyde, formaldehyde, and *m*-xylene oxidation. *J. Catal.* **1992**, *136*, 554–565. [CrossRef]
14. Ohno, T.; Tokieda, K.; Higashida, S.; Matsumura, M. Synergism between rutile and anatase TiO_2 particles in photocatalytic oxidation of naphthalene. *Appl. Catal. A* **2003**, *244*, 383–391. [CrossRef]
15. Higashimoto, S.; Kitao, N.; Yoshida, N.; Sakura, T.; Azuma, M.; Ohue, H.; Sakata, Y. Selective photocatalytic oxidation of benzyl alcohol and its derivatives into corresponding aldehydes by molecular oxygen on titanium dioxide under visible light irradiation. *J. Catal.* **2009**, *266*, 279–285. [CrossRef]
16. Ni, M.; Leung, M.K.H.; Leung, D.Y.C.; Sumathy, K. A review and recent developments in photocatalytic water-splitting using TiO_2 for hydrogen production. *Renew. Sustain. Energy Rev.* **2007**, *11*, 401–425. [CrossRef]
17. Kudo, A.; Miseki, Y. Heterogeneous photocatalyst materials for water splitting. *Chem. Soc. Rev.* **2009**, *38*, 253–278. [CrossRef] [PubMed]
18. Chen, X.; Shen, S.; Guo, L.; Mao, S.S. Semiconductor-based Photocatalytic Hydrogen Generation. *Chem. Rev.* **2010**, *110*, 6503–6570. [CrossRef] [PubMed]
19. Maeda, K.; Teramura, K.; Lu, D.; Takata, T.; Saito, N.; Inoue, Y.; Domen, K. Photocatalyst releasing hydrogen from water. *Nature* **2006**, *440*. [CrossRef] [PubMed]
20. Yang, X.; Cao, C.; Erickson, L.; Hohn, K.; Maghirang, R.; Klabunde, K. Synthesis of visible-light-active TiO_2-based photocatalysts by carbon and nitrogen doping. *J. Catal.* **2008**, *260*, 128–133. [CrossRef]
21. Han, C.; Luque, R.; Dionysiou, D.D. Facile preparation of controllable size monodisperse anatase titania nanoparticles. *Chem. Commun.* **2012**, *48*, 1860–1862. [CrossRef] [PubMed]
22. Zhu, J.; Chen, F.; Zhang, J.; Chen, H.; Anpo, M. Fe^{3+}-TiO_2 photocatalysts prepared by combining sol-gel method with hydrothermal treatment and their characterization. *J. Photochem. Photobiol. A* **2006**, *180*, 196–204. [CrossRef]
23. Wang, X.H.; Li, J.G.; Kamiyama, H.; Moriyoshi, Y.; Ishigaki, T. Wavelength-sensitive photocatalytic degradation of methyl orange in aqueous suspension over iron(III)-doped TiO_2 nanopowders under UV and visible light irradiation. *J. Phys. Chem. B* **2006**, *110*, 6804–6809. [CrossRef] [PubMed]
24. Ohno, T.; Mitsui, T.; Matsumura, M. Photocatalytic Activity of S-doped TiO_2 Photocatalyst under Visible Light. *Chem. Lett.* **2003**, *32*, 364–365. [CrossRef]
25. Virkutyte, J.; Baruwati, B.; Varma, R.S. Visible light induced photobleaching of methylene blue over melamine-doped TiO_2 nanocatalyst. *Nanoscale* **2010**, *2*, 1109–1111. [CrossRef] [PubMed]
26. Shishido, T.; Miyatake, T.; Teramura, K.; Hitomi, Y.; Yamashita, H.; Tanaka, T. Mechanism of Selective Photooxidation of Hydrocarbons over Nb_2O_5. *J. Phys. Chem.* **2009**, *113*, 18713–18718.
27. Liu, J.; Yang, S.; Wu, W.; Tian, Q.; Cui, S.; Dai, Z.; Ren, F.; Xiao, X.; Jiang, C. 3D Flowerlike α-Fe_2O_3@TiO_2 Core-Shell Nanostructures: General Synthesis and Enhanced Photocatalytic Performance. *ACS Sustain. Chem. Eng.* **2015**, *3*, 2975–2984. [CrossRef]
28. Cui, H.; Liu, Y.; Ren, W. Structure switch between α-Fe_2O_3, γ-Fe_2O_3 and Fe_3O_4 during the large scale and low temperature sol-gel synthesis of nearly monodispersed iron oxide nanoparticles. *Adv. Powder Technol.* **2013**, *24*, 93–97. [CrossRef]
29. Li, W; Yang, J.P.; Wu, Z.X.; Wang, J.X.; Li, B.; Feng, S.S.; Deng, Y.H.; Zhang, F.; Zhao, D.Y. A Versatile Kinetics-Controlled Coating Method to Construct Uniform Porous TiO_2 Shells for Multifunctional Core-Shell Structures. *J. Am. Chem. Soc.* **2012**, *134*, 11864–11867.
30. Ojeda, M.; Pineda, A.; Romero, A.A.; Barrón, V.; Luque, R. Mechanochemical Synthesis of Maghemite/Silica Nanocomposites: Advanced Materials for Aqueous Room-Temperature Catalysis. *Chem. Sustain. Chem. Energy Mater.* **2014**, *7*, 1876–1880. [CrossRef] [PubMed]

31. Colmenares, J.C.; Ouyang, W.; Ojeda, M.; Kuna, E.; Chernyayeva, O.; Lisovytskiy, D.; De, S.; Luque, R.; Balu, A.M. Mild ultrasound-assisted synthesis of TiO_2 supported on magnetic nanocomposites for selective photo-oxidation of benzyl alcohol. *Appl. Catal. B* **2015**, *183*, 107–112. [CrossRef]
32. Zhang, M.; Chen, C.; Ma, W.; Zhao, J. Visible-Light-Induced Aerobic Oxidation of Alcohols in a Coupled Photocatalytic System of Dye-Sensitized TiO_2 and TEMPO. *Angew. Chem. Int. Ed. Engl.* **2008**, *120*, 9876–9879. [CrossRef]
33. Jarry, B.; Launay, F.; Nogier, J.P.; Montouillout, V.; Gengembre, L.; Bonardet, J.L. Characterisation, acidity and catalytic activity of Ga-SBA-15 materials prepared following different synthesis procedures. *Appl. Catal. A* **2006**, *309*, 177–186. [CrossRef]
34. Ojeda, M.; Balu, A.M.; Barrón, V.; Pineda, A.; Coleto, Á.G.; Romero, A.Á.; Luque, R. Solventless mechanochemical synthesis of magnetic functionalized catalytically active mesoporous SBA-15 nanocomposites. *J. Mater. Chem. A* **2014**, *2*, 387–393. [CrossRef]
35. Qi, K.; Fei, B.; Xin, J.H. Visible light-active iron-doped anatase nanocrystallites and their self-cleaning property. *Thin Solid Films* **2011**, *519*, 2438–2444. [CrossRef]
36. Pham, M.H.; Dinh, C.T.; Vuong, G.T.; Ta, N.D.; Do, T.O. Visible light induced hydrogen generation using a hollow photocatalyst with two cocatalysts separated on two surface sides. *Phys. Chem. Chem. Phys.* **2014**, *16*, 5937–5941. [CrossRef] [PubMed]

Article

Atomic Layer Deposition of Pt Nanoparticles within the Cages of MIL-101: A Mild and Recyclable Hydrogenation Catalyst

Karen Leus [1,*], Jolien Dendooven [2], Norini Tahir [1], Ranjith K. Ramachandran [2], Maria Meledina [3], Stuart Turner [3], Gustaaf Van Tendeloo [3], Jan L. Goeman [4], Johan Van der Eycken [4], Christophe Detavernier [2] and Pascal Van Der Voort [1,*]

1 Department of Inorganic and Physical Chemistry, Center for Ordered Materials, Organometallics and Catalysis (COMOC), Ghent University, Krijgslaan 281-S3, B-9000 Ghent, Belgium; norinibinti.tahir@ugent.be
2 Department of Solid State Sciences, Conformal Coatings on Nanomaterials (CoCooN), Ghent University, Krijgslaan 281-S1, B-9000 Ghent, Belgium; jolien.dendooven@ugent.be (J.D.); ranjith.karuparambilramachandran@ugent.be (R.K.R.); christophe.detavernier@ugent.be (C.D.)
3 EMAT, University of Antwerp, Groenenborgerlaan 171, B-2020 Antwerp, Belgium; maria.meledina@uantwerpen.be (M.M.); stuart.turner@uantwerpen.be (S.T.); staf.vantendeloo@uantwerpen.be (G.V.T.)
4 Department of Organic and Macromolecular Chemistry, Laboratory for Organic and Bioorganic Synthesis, Ghent University, Krijgslaan 281-S4, B-9000 Ghent, Belgium; jan.goeman@ugent.be (J.L.G.); johan.vandereycken@ugent.be (J.V.E.)
* Correspondence: karen.leus@ugent.be (K.L.); pascal.vandervoort@ugent.be (P.V.D.V.); Tel.: +32-9-264-44-40 (K.L.); +32-9-264-44-42 (P.V.D.V.)

Academic Editors: Hermenegildo García and Sergio Navalón
Received: 1 February 2016; Accepted: 2 March 2016; Published: 9 March 2016

Abstract: We present the *in situ* synthesis of Pt nanoparticles within MIL-101-Cr (MIL = Materials Institute Lavoisier) by means of atomic layer deposition (ALD). The obtained Pt@MIL-101 materials were characterized by means of N_2 adsorption and X-ray powder diffraction (XRPD) measurements, showing that the structure of the metal organic framework was well preserved during the ALD deposition. X-ray fluorescence (XRF) and transmission electron microscopy (TEM) analysis confirmed the deposition of highly dispersed Pt nanoparticles with sizes determined by the MIL-101-Cr pore sizes and with an increased Pt loading for an increasing number of ALD cycles. The Pt@MIL-101 material was examined as catalyst in the hydrogenation of different linear and cyclic olefins at room temperature, showing full conversion for each substrate. Moreover, even under solvent free conditions, full conversion of the substrate was observed. A high concentration test has been performed showing that the Pt@MIL-101 is stable for a long reaction time without loss of activity, crystallinity and with very low Pt leaching.

Keywords: metal organic frameworks; atomic layer deposition; platinum; hydrogenation

1. Introduction

Metal Organic Frameworks (MOFs) are a class of porous crystalline materials consisting of discrete inorganic and organic secondary building units. Due to their exceptionally high porosity, pore volume, large surface area and chemical tunability and flexibility, they have already been examined in a wide range of areas such as gas storage and separations, sensing, drug delivery, ion exchange and as heterogeneous catalysts [1–3]. When used as a heterogeneous catalyst, MOFs can be examined as such or can be utilized as a support to stabilize catalytic active sites [4]. Besides the encapsulation of homogeneous complexes [5] and polyoxometalates [6], there is a growing research interest towards the embedding of nanoparticles

(NPs) into MOFs [7]. The size, shape and orientation of the NPs can be controlled by adjusting the pore size and shape of the MOFs. Moreover, the nanopores of the MOFs can be used as templates for the synthesis of monodispersed NPs. Thus far, mainly Pd [8], Au [9], Ru [10], Cu [11], Pt [12], Ni [13] and Ag [14] NPs have been incorporated into MOFs through incipient wetness impregnation, colloidal deposition, solid grinding and chemical vapor deposition.

In recent years, atomic layer deposition (ALD) has gained renewed attention as a flexible method for tailoring mesoporous materials toward specific catalytic applications [15–17]. ALD is a self-limited deposition method that is characterized by alternating exposure of the substrate to vapor-phase precursors to grow oxides, nitrides, sulfides and (noble) metals [18]. The self-limiting nature of the chemical reactions yield atomic level thickness control and excellent uniformity on complex three-dimensional supports such as mesoporous materials. In noble metal ALD, islands are often formed at the start of the ALD process instead of continuous layers. This island growth can be used advantageously to synthesize noble metal NPs on large surface area supports. Several authors have demonstrated the successful synthesis of highly dispersed Pt NPs with narrow size distributions [19–24]. Despite the unique advantages of MOFs as a scaffold in catalytic systems, noble metal ALD in MOFs has not yet been explored. The main challenge for ALD in MOFs is the slow diffusion of the chemical precursors within Ångstrom sized pores [25–27]. Therefore, Hupp and coworkers fabricated a Zr-based NU-1000 MOF with large 1D hexagonal channels (~30 Å) and successfully realized the ALD-based incorporation of acidic Al^{3+} and Zn^{2+} sites [28] and catalytically active cobalt sulfide [29]. Computational efforts provided mechanistic insight in the interaction of the ALD precursors with the MOF nodes [30]. Very recently, Jeong *et al.* reported the ALD of NiO within the framework of MIL-101-Cr (MIL = Materials Institute Lavoisier) [31]. This MIL-101 framework consists of two types of pores with inner pore diameters in the low mesoporous regime (~25–35 Å) and is thermally stable up to 300 °C. The MIL-101 framework is built up by Cr_3O-carboxylate trimers and terephthalate linkers with octahedrally coordinated metal ions binding terminal water molecules (see Figure S1) [32]. Hwang *et al.* [33] have demonstrated that these coordinated water molecules can be easily removed by a thermal treatment under vacuum at a temperature of 150 °C, creating coordinatively unsaturated sites (CUSs) which could be used, besides the Cr_3O trimers, as initial binding sites for the anchoring of nanoparticles. Because of these advantages, MIL-101-Cr was selected in this work as the MOF host for catalytically active Pt NPs synthesized by ALD. It is shown that Pt ALD results in highly dispersed, uniformly sized NPs embedded within both the small and larger MIL-101 pores. In addition, this paper reports on the catalytic properties of the Pt@MIL-101 material in the hydrogenation of different linear and cyclic olefins at room temperature.

2. Results and Discussion

2.1. *Characterization of Pt@MIL-101-Cr*

2.1.1. X-Ray Diffraction, Nitrogen Adsorption and Determination of the Pt Loading

The Pt loading of each Pt@MIL-101-Cr material was determined by means of XRF (see Table 1). As expected, an increasing number of ALD cycles resulted in a higher Pt loading. For the 40, 80 and 120 ALD cycles, a Pt loading of respectively 0.21, 0.30 and 0.35 mmol·g^{-1} was obtained. The latter sample was used for the catalytic evaluation.

Table 1. Langmuir surface area (S_{lang}) and Pt loading of the Pt@MIL-101-Cr materials (MIL = Materials Institute Lavoisier).

Sample	Pt Loading (mmol·g^{-1})	S_{lang} (m^2·g^{-1})	Pore volume (cm^3 g^{-1}) *
MIL-101-Cr	/	3614	1.52
Pt@MIL-101-Cr-40 cycles	0.21	3418	1.47
Pt@MIL-101-Cr-80 cycles	0.3	3304	1.48
Pt@MIL-101-Cr-120 cycles	0.35	3210	1.42

* After normalization for the amount of the Pt, determined at a relative pressure P/P_0 = 0.98.

Additionally, the crystallinity of the Pt@MIL-101-Cr materials was examined by means of X-ray powder diffraction (XRPD) measurements. In Figure 1, the XRPD patterns of the pristine MIL-101-Cr and the Pt@MIL-101-Cr materials obtained after the ALD deposition of the Pt nanoparticles using different cycles is presented. The XRPD pattern of each Pt@MIL-101-Cr material presents the pure phase of the non functionalized MIL-101-Cr. This explicitly shows that the framework integrity of the parent MOF was well preserved during the ALD deposition process, despite the use of ozone as reactant. Nitrogen sorption measurements were carried out to determine the Langmuir surface area of the pristine MIL-101-Cr and Pt@MIL-101-Cr materials (see Table 1 and Figure S2 for the nitrogen adsorption isotherms). The MIL-101-Cr has a Langmuir surface area of 3614 m^2/g, which is significantly higher than the value usually reported in literature because of an extra activation step carried out to remove the free organic linker [34,35]. Only the group of Férey reported a higher Langmuir surface area of approximately 5900 m^2/g by adding hydrogen fluoride (HF) to the synthesis of the framework [36]. No significant change in the Langmuir surface area is observed for the Pt@MIL-101-Cr materials obtained after 40 cycles and even after 80 ALD cycles. In addition, inspecting the capillary condensation step in Figure S2, it is clear that there is no obvious pore size reduction upon increasing the number of ALD cycles which is in contrast to the work of Snurr, Hupp and Farha [28,37], but, in these cases, metal oxides were prepared by cycling a metal precursor and water in the MOF framework. These adsorption data corroborate the finding that Pt (as a zerovalent metal) is formed as nanoparticles inside the pores, rather than by a layer by layer deposition on the walls, which would result in a gradual pore mouth and cage size reduction. This is further corroborated by tomography and transmission electron microscopy (TEM) data.

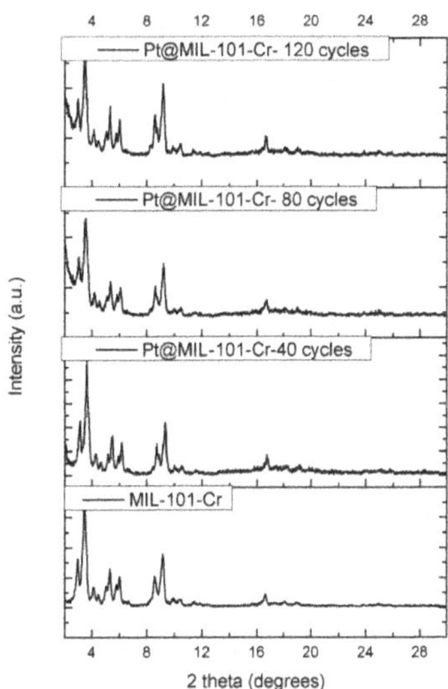

Figure 1. X-ray powder diffraction (XRPD) patterns of MIL-101-Cr and the obtained Pt@MIL-101-Cr materials (MIL = Materials Institute Lavoisier).

2.1.2. TEM Measurements

In order to investigate the Pt loading in the Pt@MIL-101-Cr, (high angle) annular dark field scanning transmission electron microscopy measurements ((HA)ADF-STEM) were carried out on the Pt@MIL-101-Cr-120 cycles sample. As MOFs are known to be extremely sensitive to the electron beam, the electron dose, dwell time and the image magnification were optimized in order to acquire images of the intact MIL-101-Cr framework [38].

The MIL-101-Cr crystals in the Pt@MIL-101-Cr-120 cycles sample demonstrate a typical truncated octahedral morphology, with predominant {111} facets and {100} truncation (Figure 2a). It is clear from the high magnification ADF-STEM images (bottom row in Figure 2) that the MIL-101-Cr crystals maintain their initial crystallinity after the ALD deposition of Pt nanoparticles. The bright contrast features in the images correspond to the heavy Pt nanoparticles which are evenly dispersed in the MIL-101-Cr crystals. The Pt nanoparticle size matches well with the pore diameter of the MIL-101-Cr framework, indicating that they are likely embedded within the pores of the MIL-101-Cr framework. To completely fill the smaller cages with Pt, around 900 Pt atoms are needed, and, in the case of the bigger cages, ~1400 atoms, which is in accordance with the observed Pt NP size of ~2–3 nm. However, the HA(ADF) images are only 2D projections of 3D objects. The direct method to determine the 3D position of the nanoparticles is electron tomography which has been performed on the ALD-loaded Pt@MIL-101 in our previous work. An additional electron tomography series was acquired in this study (see Figure S3 and Movie M.1) on a heavily Pt loaded MIL-101-Cr crystal, which unambiguously demonstrates that the ALD loading of Pt NP into the MIL-101-Cr frameworks leads to the embedding of nanoparticles inside the cages of the MOF host [38]. However, it is also clear from this and our previous study that some of the Pt is remaining at the surface, mostly in the form of larger Pt chunks.

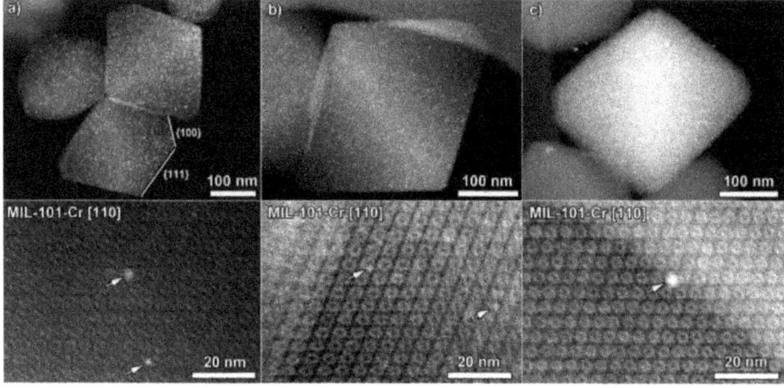

Figure 2. (High angle) annular dark field scanning transmission electron microscopy measurements ((HA)ADF-STEM) (**top row**) and ADF-STEM (**bottom row**) images. (**a**) Fresh Pt@MIL-101-Cr-120 cycles; (**b**) Pt@MIL-101-Cr-120 cycles after run 1; (**c**) Pt@MIL-101-Cr-120 cycles after the high concentration run. The white arrows point to Pt nanoparticles with similar diameters to the MIL-101 framework pores.

2.2. Catalytic Results

A number of reports have already demonstrated the potential of Pt nanoparticles as hydrogenation catalysts. Within this regard, Pt nanoparticles have been immobilized on different supports like carbon nanotubes, silica based materials and MOFs [39,40]. While, in this study, cyclic and linear olefins were utilized as substrates to examine the catalytic performance of Pt@MIL-101-Cr, other studies have used this MOF for the hydrogenation of nitroarenes [41,42], cinnamaldehyde [43] and for the assymetric hydrogenation of α-ketoesters [44]. Additionally, besides these liquid phase based hydrogenation reactions, gas phase olefin hydrogenation reactions have been reported for Pt@MOF catalysts [45,46]. In Table 2, an overview is presented of the investigated substrates employing Pt@MIL-101-Cr-120 cycles as catalyst, compared to some other studies that used Pt@MOF catalysts, while in Figure S4 the conversion patterns using Pt@MIL-101-Cr-120 cycles as catalyst are shown. In Table 3, the TON, TOF and Pt leaching is presented. As can be seen from Table 2, the Pt@MIL-101-Cr-120 cycles exhibits approximately full conversion in the hydrogenation of each examined substrate. For 1-octene full conversion was observed after only 30 min of reaction (entry 5), whereas for styrene (entry 6) 97% conversion was noted after 3 h with the formation of respectively n-Octane and ethyl benzene. However, it is important to note that the blank reactions for the latter substrates already showed a high converison. More specifically, a conversion of 37% of 1-octene was seen after 30 min of reaction whereas, for styrene, 50% was already converted after 3 h of reaction in the absence of the catalyst. For the other examined substrates, cyclohexene and cyclooctene, the conversions obtained for the blank reactions were significantly lower with only 11% cyclohexene conversion after 2 h of reaction time and no conversion of cyclooctene under these reaction conditions. In the presence of the catalyst, 94% of cyclooctene was converted after 6 h (entry 7), whereas, for cyclohexene, 98% of conversion was noted after just two hours of reaction (entry 8). The latter substrate can also be converted under solvent free conditions (entry 9). Full conversion was noted after 20 h of reaction.

Table 2. Comparison of the catalytic activity of Pt@MIL-101-120 cycles with other Pt based Metal Organic Frameworks (MOF) catalysts in the hydrogenation of cyclic and linear olefins.

Entry	Catalyst	Substrate	Reaction Conditions	Reaction Time	Conversion	Main Product	Reference
1	Pt@MIL-101	1-octene	35 °C, solvent free at 1.5 bar of H_2	6 h	>99%	n-Octane	[47]
2	Pt@ZIF-8	1-hexene	RT, ethanol at 1 bar of H_2	24 h	>95%	n-Hexane	[48]
3	Pt@ZIF-8	cyclooctene	RT, ethanol at 1 bar of H_2	24 h	2.7%	Cyclooctane	[48]
4	Pt-Ni frame@ Ni-MOF-74	Styrene	30 °C, THF at 1 bar of H_2	3 h	>99%	/	[49]
5	Pt@MIL-101	1-octene	RT, ethanol at 6 bar of H_2	30 min	>99%	n-Octane	this work
6	Pt@MIL-101	Styrene	RT, ethanol at 6 bar of H_2	3h	>97%	Ethyl benzene	this work
7	Pt@MIL-101	cyclooctene	RT, ethanol at 6 bar of H_2	6h	>94%	Cyclooctane	this work
8	Pt@MIL-101	cyclohexene	RT, ethanol at 6 bar of H_2	2h	>98%	Cyclohexane	this work
9	Pt@MIL-101	cyclohexene	60 °C, solvent free at 6 bar of H_2	20h	>99%	Cyclohexane	this work

Table 3. The turnover number (TON), turnover frequency (TOF) and leaching percentage for each examined substrate using Pt@MIL-101-Cr-120 cycles as catalyst.

Substrate	TON	TOF (min^{-1})	Reaction Time	Leaching of Pt (%)
1-Octene	497	16.6	30 min	<0.05 *
Styrene	482.7	3.7	3h	0.89
Cyclohexene	490	4.4	2h	0.32
Cyclooctene	468	1.93	6h	<0.05 *

* Below detection limit. The TON number was determined at the end of the reaction while the TOF number was determined after 30 min of catalysis.

Additionally, the Pt@MIL-101-Cr-120 cycles' catalyst was compared with other Pt@MOF based heterogeneous catalysts for the hydrogenation of linear and cyclic olefins. Although it is difficult to give an objective comparison, as different catalytic conditions were used in these tests, it can be seen from Table 2 that each Pt based catalyst, including the Pt@MIL-101-Cr-120 cycles, exhibits a good catalytic performance in the hydrogenation of alkenes except for the Pt@ZIF-8 (ZIF = zeolitic imidazolate framework) in the hydrogenation of cyclooctene. For the substrate 1-hexene, using Pt@ZIF-8, 95% of conversion in 24 h was obtained with no side product formation, whereas almost no activity was seen in the hydrogenation of cyclooctene (2.7%) (entry 2 and 3). The authors adressed this difference in activity to the difference in size of the examined substates, as cyclooctene has a molecular width of 5.7 Å which exceeds the size of the apertures of ZIF-8 (3.4 Å) [48]. The pore aperture in MIL-101-Cr is significantly larger (12 and 15 Å), as MIL-101-Cr contains two types of cages having a diameter of respectively 29 Å and 34 Å. The enhanced catalytic performance of the Pt@MIL-101-Cr-120 cycles catalyst in comparison to the reported Pt@ZIF-8 for the hydrogenation of cyclooctene can be assigned to this difference in pore aperture. For cyclohexene, styrene and cyclooctene the turnover frequency (TOF) is respectively 4.4 min^{-1}, 3.7 min^{-1} and 1.93 min^{-1}, as can be expected as the kinetics slow down as the molecules become larger. Furthermore, the room temperature based conversion of styrene in this work, in a very recent study of Li *et al.* [49], the latter substrate was fully converted at the same reaction time (3 h) under a H_2 pressure of 1 bar through use of a bimetallic Pt-Ni frame@MOF-74, but no product distribution was presented (entry 4).

2.3. *Reusability and Stability Tests*

To examine the reusability of the Pt@MIL-101-Cr-120 cycles catalyst, a high concentration run was carried out in which 10 times more substrate was added in comparison to the previous catalytic experiments, without changing the catalyst loading. This procedure is employed when little catalyst is available, as repeated filtration steps result in cumulative catalyst losses. In addition, 250 Mmol of cyclooctene was added into the Parr reactor and the reaction was monitored until full conversion was obtained. As can be seen from Figure S5, nearly full conversion was reached after approximately 168 h of reaction. This observation demonstrates that the Pt@MIL-101-Cr-120 cycles catalyst does not lose its activity nor becomes deactivated during many turnovers. Additionally, during this high concentration run, only a negligible amount of Pt NPs was leached out: 0.81% of Pt was leached from the Pt@MIL-101-Cr-120 cycles. The turnover number TON (determined at the end of the reaction) and TOF (determined after 2 h of catalysis) number for this concentrated run is respectively 4859 and 108 h^{-1}. Comparison of the XRPD patterns of the Pt@MIL-101-Cr-120 cycles before and after catalysis clearly shows that no changes are observed in the XRPD pattern of the Pt@MIL-101-Cr-120 cycles after catalysis when compared to the pristine MOF, even after the high concentration run (see Figure 3). The latter observation shows that the framework integrity of the MOF was preserved.

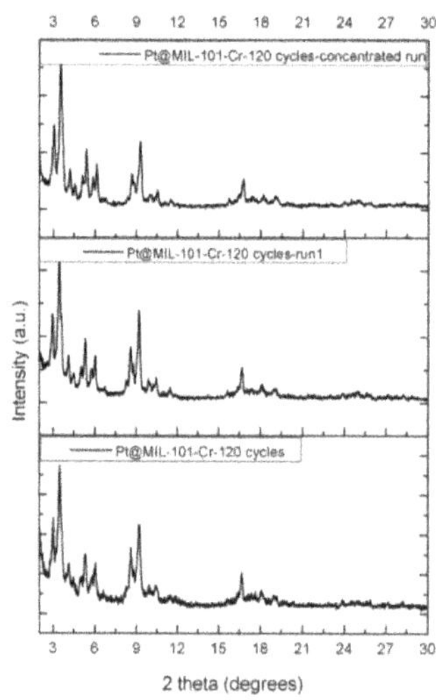

Figure 3. XRPD pattern of the pristine catalyst and after the first run and the concentrated run.

Additionally, in Figure S6, the nitrogen adsorption isotherms are presented for the Pt@MIL-101 material before and after catalysis. From this figure, one can see that the Langmuir surface area slightly decreased after catalysis, which is probably due to a partial clogging of the pores during the catalytic testing. The fresh catalyst has a Langmuir surface area of 3210 m^2/g, the Pt@MIL-101-Cr-120 cycles-run 1 and Pt@MIL-101-Cr-120 cycles-concentrated run have a Langmuir surface area of 2650 and 2700 m^2/g respectively. Moreover, annular dark field scanning transmission electron microscopy measurements were carried out on the Pt@MIL-101-Cr-120 cycles after run 1 (Figure 2b) and after the high concentration run (Figure 2c). It can be seen from these images that the morphology of the MIL-101 particles remains similar to these of the fresh Pt@MIL-101-Cr-120 cycles sample and that the MIL-101 framework retains its crystalline nature. Additionally, no significant agglomeration of the Pt nanoparticles is observed, even after the high concentration run.

3. Experimental Section

3.1. *Materials and Methods*

All chemicals were purchased from Sigma Aldrich (Diegem, Belgium) or TCI Europe (Zwijndrecht, Belgium) and used without further purification. Nitrogen adsorption experiments were carried out at −196 °C using a Belsorp-mini II gas analyzer (Rubotherm, Bochum, Germany). Prior to analysis, the samples were dried under vacuum at 90 °C to remove adsorbed water. XRPD patterns were collected on a ARL X'TRA X-ray diffractometer (Thermo Fisher Scientific, Erembodegem, Belgium) with Cu Ka radiation of 0.15418 nm wavelength and a solid state detector. XRF measurements were performed on a NEX CG from Rigaku (Addspex, Abcoude, the Netherlands) using a Mo-X-ray source (Addspex, Abcoude, the Netherlands) Elemental analyses was conducted using a Vista-MPX CCD Simultaneous Inductively Coupled Plasma-Optical Emission Spectrometer (ICP-OES) (Waltham, MA, USA). HAADF-STEM and ADF-STEM imaging was carried out on a FEI Tecnai Osiris microscope (Hillsboro, Oregon, USA), operated at 200 kV. The convergence semi-angle used was 10 mrad, the inner ADF detection angle was 14 mrad, the inner HAADF-STEM detection angle was 50 mrad.

3.2. *Catalytic Setup*

In each catalytic test, the Parr reactor was loaded with 70.0 mL ethanol and 2.84 mL of dodecane used respectively as solvent and internal standard. The examined substrates in this study are 1-octene, styrene, cyclooctene and cyclohexene (25 mmol). The Pt@MIL-101-Cr-120 cycles was used as the catalyst. For each examined substrate, the same loading of active sites was employed. More specifically in every catalytic test, 0.05 mmol Pt sites were used which give rise to a molar ratio of substrate: catalyst of 500:1. The TON number was calculated by dividing the mmol obtained product by the number of active sites while the TOF number was determined by dividing the TON number by the reaction time (expressed in minutes). All the catalytic tests were performed at room temperature and at a pressure of 6 bar H_2. During each test, aliquots were gradually taken out of the mixture and subsequently analyzed by means of gas chromatography (GC) using a split injection (ratio 1:17) on a Hewlett Packard 5890 Series II GC with TCD detection (Santa Clara, CA, USA). The capillary column used was a Restek XTI-5 column (Bellefonte, PA, USA) with a length of 30 m, an internal diameter of 0.25 mm and a film thickness of 0.25 μm. H_2 was used as carrier gas under constant flow conditions (1.4 mL/min).The fresh catalyst was activated under vacuum at 90 °C overnight prior to catalysis. After each catalytic run, the catalyst was recovered by filtration, washed with acetone, and dried at 90 °C overnight under vacuum.

3.3. *Synthesis of MIL-101-Cr and Pt@MIL-101-Cr*

The MIL-101-Cr was synthesized according to a slightly modified procedure of Edler *et al.* [34]. Typically, 0.6645 g of terephthalic acid was mixed with 1.6084 g of $Cr(NO_3)_3 \cdot 9H_2O$ and 20 mL of destilled water. The mixture was transferred into a Teflon-lined autoclave, sealed and heated in 2 h to 210 °C at which it was hold for 8 h. After cooling down to room temperature, the MIL-101*as* was filtered, washed with acetone and stirred in dimethylformamide for 24 h at 60 °C to remove unreacted terephthalic acid. Thereafter, the MIL-101 was stirred in 1M HCl for 12 h at room temperature, filtered and dried under vacuum at 90 degrees overnight. The deposition of the Pt nanoparticles was achieved by means of ALD. Pt ALD on the powder sample was performed at 200 °C using (methylcyclopentadienyl)-trimethylplatinum [$MeCpPtMe_3$] as Pt source and O_3 as reactant [50]. All the depositions were conducted in a home built experimental cold-wall ALD chamber connected through a gate valve to a turbo pump backed up by a rotary pump. A second gate valve was installed for pre-evacuation of the chamber via a bypass line to the rotary pump. The powder sample was loaded in a molybdenum sample cup which was then transferred into the ALD reactor through the load-lock and was placed on a heater block. After loading into the reactor, the powder sample was allowed to outgas and thermally equilibrate for at least 1 h under vacuum. The solid $MeCpPtMe_3$ precursor (99% Strem Chemicals), kept in a stainless steel container, was heated above its melting point (30 °C), and the delivery line to the chamber was heated to 60 °C. Argon was used as a carrier gas for the Pt precursor. O_3 was produced from a pure O_2 flow with an OzoneLab™ OL100 ozone generator (Ozone Services, Burton, BC, Canada), resulting in an O_3 concentration of 175 µg/mL. A static exposure mode was applied during both ALD half-cycles [1,4]. The pulse time of the $MeCpPtMe_3$ precursor was 10 s, after which the valves to the pumping system were kept closed for another 20 s, resulting in a total exposure time of 30 s. The same pulse time and exposure time was used for the O_3 also. The effect of the exposure times on the Pt loading was not studied in detail. Nevertheless, these exposure times were found to be large enough to ensure penetration deep into the MOF crystals. During the precursor and reactant exposures, the pressure in the chamber increased to ca. 5×10^{-1} mbar and 1 mbar, respectively. In between the two exposures, the valve to the rotary pump was first opened for 10 s and then the valve to the turbo pump was opened for another 50 s to reach the base pressure.

4. Conclusions

Pt NPs were synthesized *in situ* within MIL-101-Cr by means of ALD, enabling the varying of Pt loading by changing the number of ALD cycles. Highly dispersed Pt NPs were obtained with sizes determined by the pore sizes of the MOF host. The Pt@MIL-101 materials maintained their porosity and crystallinity during the synthesis of the Pt NPs and during the catalytic hydrogenation of cyclic and linear olefins. Full conversion for every substrate was obtained using Pt@MIL-101 as catalyst under mild reaction conditions. Moreover, even under solvent free conditions, full conversion was shown with negligible leaching of Pt. Stability tests have demonstrated that the Pt@MIL-101 catalyst is stable for a long reaction time without loss in crystallinity or agglomeration of the Pt NPs, and with a high TOF and TON number.

Supplementary Materials: The following are available online at http://www.mdpi.com/2079-4991/6/3/45/s1.

Acknowledgments: Karen Leus acknowledges the financial support from the Ghent University "Bijzonder Onderzoeksfonds" BOF post-doctoral Grant 01P06813T and UGent "Geconcentreeerde Onderzoekacties" GOA Grant 01G00710. Jolien Dendooven and Stuart Turner gratefully acknowledges the "Fonds Wetenschappelijk Onderzoek" FWO Vlaanderen for a post-doctoral scholarship. Christophe Detavernier thanks the FWO Vlaanderen, BOF-UGent (GOA 01G01513) and the Hercules Foundation (AUGE/09/014) for financial support. The Titan microscope used for this investigation was partially funded by the Hercules foundation of the Flemish government. This work was supported by the "Belgian Interuniversitaire Attractie Pool-Pôle d'Attraction Interuniversitaire" IAP-PAI network.

Author Contributions: Christophe Detavernier and Pascal Van Der Voort had the original idea to anchor nanoparticles in the MIL-101 by means of ALD. Karen Leus and Norini Tahir carried out the synthesis of the MIL-101 material, characterization and catalytic testing of the Pt@MIL-101 materials whereas Jan L. Goeman and Johan Van der Eycken performed the GC analysis. Jolien Dendooven and Ranjith K. Ramachandran have done the ALD depositions. Maria Meledina, Stuart Turner and Gustaaf Van Tendeloo have performed the HAADF-STEM and ADF-STEM measurements. All authors read and approved the final manuscript.

Conflicts of Interest: The authors declare no conflict of interest.

References

1. Cai, W.; Chu, C.C.; Liu, G.; Wang, Y.X.J. Metal-Organic Framework-Based Nanomedicine Platforms for Drug Delivery and Molecular Imaging. *Small* **2015**, *11*, 4806–4822. [CrossRef] [PubMed]

2. Leus, K.; Liu, Y.Y.; van der Voort, P. Metal-Organic Frameworks as Selective or Chiral Oxidation Catalysts. *Catal. Rev.* **2014**, *56*, 1–56. [CrossRef]
3. Barea, E.; Montoro, C.; Navarro, J.A.R. Toxic gas removal metal-organic frameworks for the capture and degradation of toxic gases and vapours. *Chem. Soc. Rev.* **2014**, *43*, 5419–5430. [CrossRef] [PubMed]
4. Leus, K.; Liu, Y.Y.; Meledina, M.; Turner, S.; van Tendeloo, G.; van der Voort, P. A Mo-VI grafted Metal Organic Framework: Synthesis, characterization and catalytic investigations. *J. Catal.* **2014**, *316*, 201–209. [CrossRef]
5. Bogaerts, T.; van Yperen-De Deyne, A.; Liu, Y.Y.; Lynen, F.; van Speybroeck, V.; van der Voort, P. Mn-salen@MIL-101(Al): A heterogeneous, enantioselective catalyst synthesized using a 'bottle around the ship' approach. *Chem. Commun.* **2013**, *49*, 8021–8023. [CrossRef] [PubMed]
6. Juan-Alcaniz, J.; Ramos-Fernandez, E.V.; Lafont, U.; Gascon, J.; Kapteijn, F. Building MOF bottles around phosphotungstic acid ships: One–pot synthesis of bi-functional polyoxometalate-MIL-101 catalysts. *J. Catal.* **2010**, *269*, 229–241. [CrossRef]
7. Meilikhov, M.; Yusenko, K.; Esken, D.; Turner, S.; van Tendeloo, G.; Fischer, R.A. Metals@MOFs-Loading MOFs with Metal Nanoparticles for Hybrid Functions. *Eur. J. Inorg. Chem.* **2010**, *2010*, 3701–3714. [CrossRef]
8. Luz, I.; Rosler, C.; Epp, K.; Xamena, F.X.L.I.; Fischer, R.A. Pd@UiO-66-Type MOFs Prepared by Chemical Vapor Infiltration as Shape-Selective Hydrogenation Catalysts. *Eur. J. Inorg. Chem.* **2015**, *2015*, 3904–3912. [CrossRef]
9. Leus, K.; Concepcion, P.; Vandichel, M.; Meledina, M.; Grirrane, A.; Esquivel, D.; Turner, S.; Poelman, D.; Waroquier, M.; van Speybroeck, V.; *et al.* Au@UiO-66: a base free oxidation catalyst. *RSC Adv.* **2015**, *5*, 22334–22342. [CrossRef]
10. Schroeder, F.; Esken, D.; Cokoja, M.; van den Berg, M.W.E.; Lebedev, O.I.; van Tendeloo, G.; Walaszek, B.; Buntkowsky, G.; Limbach, H.H.; Chaudret, B.; *et al.* Ruthenium nanoparticles inside porous [$Zn_4O(bdc)_3$] by hydrogenolysis of adsorbed [Ru(cod)(cot)]: A solid-state reference system for surfactant-stabilized ruthenium colloids. *J. Am. Chem. Soc.* **2008**, *130*, 6119–6130. [CrossRef] [PubMed]
11. Juan-Alcaniz, J.; Ferrando-Soria, J.; Luz, I.; Serra-Crespo, P.; Skupien, E.; Santos, V.P.; Pardo, E.; Xamena, F.X.L.I.; Kapteijn, F.; Gascon, J. The oxamate route, a versatile post-functionalization for metal incorporation in MIL-101(Cr): Catalytic applications of Cu, Pd and Au. *J. Catal.* **2013**, *307*, 295–304. [CrossRef]
12. Xu, Z.D.; Yang, L.Z.; Xu, C.L. Pt@UiO-66 Heterostructures for Highly Selective Detection of Hydrogen Peroxide with an Extended Linear Range. *Anal. Chem.* **2015**, *87*, 3438–3444. [CrossRef] [PubMed]
13. Zhao, H.H.; Song, H.L.; Chou, L.J. Nickel nanoparticles supported on MOF-5: Synthesis and catalytic hydrogenation properties. *Inorg. Chem. Commun.* **2012**, *15*, 261–265. [CrossRef]
14. Abdelhameed, R.M.; Simoes, M.M.Q.; Silva, A.M.S.; Rocha, J. Enhanced Photocatalytic Activity of MIL-125 by Post-Synthetic Modification with Cr-III and Ag Nanoparticles. *Chem. Eur. J.* **2015**, *21*, 11072–11081. [CrossRef] [PubMed]
15. O'Neill, B.J.; Jackson, D.H.K.; Lee, J.; Canlas, C.; Stair, P.C.; Marshall, C.L.; Elam, J.W.; Kuech, T.F.; Dumesic, J.A.; Huber, G.W. Catalyst Design with Atomic Layer Deposition. *ACS Catal.* **2015**, *5*, 1804–1825. [CrossRef]
16. Detavernier, C.; Dendooven, J.; Sree, S.P.; Ludwig, K.F.; Martens, J.A. Tailoring nanoporous materials by atomic layer deposition. *Chem. Soc. Rev.* **2011**, *40*, 5242–5253. [CrossRef] [PubMed]
17. Dendooven, J. *Atomically-Precise Methods for Synthesis of Solid Catalysts*; Hermans, S., Visart de Bocarme, T., Eds.; RSC: Cambridge, MA, USA, 2015; Volume 1, p. 167.
18. Miikkulainen, V.; Leskela, M.; Ritala, M.; Puurunen, R.L. Crystallinity of inorganic films grown by atomic layer deposition: Overview and general trends. *J. Appl. Phys.* **2013**, *113*. [CrossRef]
19. King, J.S.; Wittstock, A.; Biener, J.; Kucheyev, S.O.; Wang, Y.M.; Baumann, T.F.; Giri, S.K.; Hamza, A.V.; Baeumer, M.; Bent, S.F. Ultralow loading of Pt nanocatalysts prepared by atomic layer deposition on carbon aerogels. *Nano Lett.* **2008**, *8*, 2405–2409. [CrossRef] [PubMed]
20. Gould, T.D.; Lubers, A.M.; Corpuz, A.R.; Weimer, A.W.; Falconer, J.L.; Medlin, J.W. Controlling Nanoscale Properties of Supported Platinum Catalysts through Atomic Layer Deposition. *ACS Catal.* **2015**, *5*, 1344–1352. [CrossRef]
21. Goulas, A.; van Ommen, J.R. Atomic layer deposition of platinum clusters on titania nanoparticles at atmospheric pressure. *J. Mater. Chem. A* **2013**, *1*, 4647–4650. [CrossRef]

22. Enterkin, J.A.; Setthapun, W.; Elam, J.W.; Christensen, S.T.; Rabuffetti, F.A.; Marks, L.D.; Stair, P.C.; Poeppelmeier, K.R.; Marshall, C.L. Propane Oxidation over Pt/$SrTiO_3$ Nanocuboids. *ACS Catal.* **2011**, *1*, 629–635. [CrossRef]
23. Zhou, Y.; King, D.M.; Liang, X.H.; Li, J.H.; Weimer, A.W. Optimal preparation of Pt/TiO_2 photocatalysts using atomic layer deposition. *Appl. Catal. B* **2010**, *101*, 54–60. [CrossRef]
24. Li, J.H.; Liang, X.H.; King, D.M.; Jiang, Y.B.; Weimer, A.W. Highly dispersed Pt nanoparticle catalyst prepared by atomic layer deposition. *Appl. Catal. B* **2010**, *97*, 220–226. [CrossRef]
25. Dendooven, J.; Devloo-Casier, K.; Ide, M.; Grandfield, K.; Kurttepeli, M.; Ludwig, K.F.; Bals, S.; van der Voort, P.; Detavernier, C. Atomic layer deposition-based tuning of the pore size in mesoporous thin films studied by *in situ* grazing incidence small angle X-ray scattering. *Nanoscale* **2014**, *6*, 14991–14998. [CrossRef] [PubMed]
26. Sree, S.P.; Dendooven, J.; Jammaer, J.; Masschaele, K.; Deduytsche, D.; D'Haen, J.; Kirschhock, C.E.A.; Martens, J.A.; Detavernier, C. Anisotropic Atomic Layer Deposition Profiles of TiO_2 in Hierarchical Silica Material with Multiple Porosity. *Chem. Mater.* **2012**, *24*, 2775–2780. [CrossRef]
27. Dendooven, J.; Goris, B.; Devloo-Casier, K.; Levrau, E.; Biermans, E.; Baklanov, M.R.; Ludwig, K.F.; van der Voort, P.; Bals, S.; Detavernier, C. Tuning the Pore Size of Ink-Bottle Mesopores by Atomic Layer Deposition. *Chem. Mater.* **2012**, *24*, 1992–1994. [CrossRef]
28. Mondloch, J.E.; Bury, W.; Fairen-Jimenez, D.; Kwon, S.; DeMarco, E.J.; Weston, M.H.; Sarjeant, A.A.; Nguyen, S.T.; Stair, P.C.; Snurr, R.Q.; *et al.* Vapor-Phase Metalation by Atomic Layer Deposition in a Metal-Organic Framework. *J. Am. Chem. Soc.* **2013**, *135*, 10294–10297. [CrossRef] [PubMed]
29. Peters, A.W.; Li, Z.Y.; Farha, O.K.; Hupp, J.T. Atomically Precise Growth of Catalytically Active Cobalt Sulfide on Flat Surfaces and within a Metal-Organic Framework via Atomic Layer Deposition. *ACS Nano* **2015**, *9*, 8484–8490. [CrossRef] [PubMed]
30. Kim, I.S.; Borycz, J.; Platero-Prats, A.E.; Tussupbayev, S.; Wang, T.C.; Farha, O.K.; Hupp, J.T.; Gagliardi, L.; Chapman, K.W.; Cramer, C.J.; *et al.* Targeted Single-Site MOF Node Modification: Trivalent Metal Loading via Atomic Layer Deposition. *Chem. Mater.* **2015**, *27*, 4772–4778. [CrossRef]
31. Jeong, M.-G.; Kim, D.H.; Lee, S.-K.; Lee, J.H.; Han, S.W.; Park, E.J.; Cychosz, K.A.; Thommes, M.; Hwang, Y.K.; Chang, J.-S.; *et al.* Decoration of the internal structure of mesoporous chromium terephthalate MIL-101 with NiO using atomic layer deposition. *Microporous Mesoporous Mater.* **2016**, *221*, 101–107. [CrossRef]
32. Maksimchuk, N.V.; Zalomaeva, O.V.; Skobelev, I.Y.; Kovalenko, K.A.; Fedin, V.P.; Kholdeeva, O.A. Metal-organic frameworks of the MIL-101 family as heterogeneous single-site catalysts. *Proc. R. Soc.* **2012**, *468*, 2017–2034. [CrossRef]
33. Hwang, Y.K.; Hong, D.Y.; Chang, J.S.; Seo, H.; Yoon, M.; Kim, J.; Jhung, S.H.; Serre, C.; Ferey, G. Selective sulfoxidation of aryl sulfides by coordinatively unsaturated metal centers in chromium carboxylate MIL-101. *Appl. Catal.* **2009**, *358*, 249–253. [CrossRef]
34. Jiang, D.M.; Burrows, A.D.; Edler, K.J. Size-controlled synthesis of MIL-101(Cr) nanoparticles with enhanced selectivity for CO_2 over N_2. *Crystengcomm* **2011**, *13*, 6916–6919. [CrossRef]
35. Leus, K.; Bogaerts, T.; de Decker, J.; Depauw, H.; Hendrickx, K.; Vrielinck, H.; van Speybroeck, V.; van der Voort, P. Systematic Study of the Chemical and Hydrothermal Stability of Selected "Stable" Metal Organic Frameworks. *Microporous Mesoporous Mater.* **2016**, *226*, 110–116. [CrossRef]
36. Ferey, G.; Mellot-Draznieks, C.; Serre, C.; Millange, F.; Dutour, J.; Surble, S.; Margiolaki, I. A chromium terephthalate-based solid with unusally large pore volumes and surface area. *Science* **2005**, *309*, 2040–2042. [CrossRef] [PubMed]
37. Deria, P.; Mondloch, J.E.; Tylianakis, E.; Ghosh, P.; Bury, W.; Snurr, R.Q.; Hupp, J.T.; Farha, O.K. Perfluoroalkane Functionalization of NU-1000 via Solvent-Assisted Ligand Incorporation: Synthesis and CO_2 Adsorption Studies. *J. Am. Chem. Soc.* **2013**, *135*, 16801–16804. [CrossRef] [PubMed]
38. Meledina, M.; Turner, S.; Filippousi, M.; Leus, K.; Lobato, I.; Ramachandran, R.K.; Dendooven, J.; Detavernier, C.; van der Voort, P.; van Tendeloo, G. Direct Imaging of ALD Deposited Pt Nanoclusters inside the Giant Pores of MIL-101. *Part. Part. Syst. Charact.* **2016**. [CrossRef]
39. Zhang, S.; Shao, Y.Y.; Yin, G.P.; Lin, Y.H. Carbon nanotubes decorated with Pt nanoparticles via electrostatic self-assembly: a highly active oxygen reduction electrocatalyst. *J. Mater. Chem.* **2010**, *20*, 2826–2830. [CrossRef]

40. Joo, S.H.; Park, J.Y.; Tsung, C.K.; Yamada, Y.; Yang, P.D.; Somorjai, G.A. Thermally stable Pt/mesoporous silica core-shell nanocatalysts for high-temperature reactions. *Nat. Mater.* **2009**, *8*, 126–131. [CrossRef] [PubMed]
41. Du, W.C.; Chen, G.Z.; Nie, R.F.; Li, Y.W.; Hou, Z.Y. Highly dispersed Pt in MIL-101: An efficient catalyst for the hydrogenation of nitroarenes. *Catal. Commun.* **2013**, *41*, 56–59. [CrossRef]
42. Pan, H.Y.; Li, X.H.; Yu, Y.; Li, J.R.; Hu, J.; Guan, Y.J.; Wu, P. Pt nanoparticles entrapped in mesoporous metal-organic frameworks MIL-101 as an efficient catalyst for liquid-phase hydrogenation of benzaldehydes and nitrobenzenes. *J. Mol. Catal.* **2015**, *399*, 1–9. [CrossRef]
43. Liu, H.L.; Li, Z.; Li, Y.W. Chemoselective Hydrogenation of Cinnamaldehyde over a Pt-Lewis Acid Collaborative Catalyst under Ambient Conditions. *Ind. Eng. Chem. Res.* **2015**, *54*, 1487–1497. [CrossRef]
44. Pan, H.Y.; Li, X.H.; Zhang, D.M.; Guan, Y.J.; Wu, P. Pt nanoparticles entrapped in mesoporous metal-organic frameworks MIL-101 as an efficient and recyclable catalyst for the assymetric hydrogenation of alpha-ketoesters. *J. Mol. Catal.* **2013**, *377*, 108–114. [CrossRef]
45. Guo, Z.Y.; Xiao, C.X.; Maligal-Ganesh, R.V.; Zhou, L.; Goh, T.W.; Li, X.L.; Tesfagaber, D.; Thiel, A.; Huang, W.Y. Pt Nanoclusters Confined within Metal Organic Framework Cavities for Chemoselective Cinnamaldehyde Hydrogenation. *ACS Catal.* **2014**, *4*, 1340–1348. [CrossRef]
46. Ramos-Fernandez, E.V.; Pieters, C.; van der Linden, B.; Juan-Alcaniz, J.; Serra-Crespo, P.; Verhoeven, M.W.G.M.; Niemantsverdriet, H.; Gascon, J.; Kapteijn, F. Highly dispersed platinum in metal organic framework NH_2-MIL-101(Al) containing phosphotungstic acid- Characterization and catalytic performance. *J. Catal.* **2012**, *289*, 42–52. [CrossRef]
47. Khajavi, H.; Stil, H.A.; Kuipers, H.P.C.E.; Gascon, J.; Kapteijn, F. Shape and Transition State Selective Hydrogenations Using Egg-Shell Pt-MIL-101(Cr) Catalyst. *ACS Catal.* **2013**, *3*, 2617–2626. [CrossRef]
48. Wang, P.; Zhao, J.; Li, X.B.; Yang, Y.; Yang, Q.H.; Li, C. Assembly of ZIF nanostructures around free Pt nanoparticles: Efficient size-selective catalysts for hydrogenation of alkenes under mild conditions. *Chem. Commun.* **2013**, *49*, 3330–3332. [CrossRef] [PubMed]
49. Li, Z.; Yu, R.; Huang, J.L.; Shi, Y.S.; Zhang, D.Y.; Zhong, X.Y.; Wang, D.S.; Wu, Y.E.; Li, Y.D. Platinum-nickel frame within metal-organic framework fabricated in situ for hydrogen enrichment and molecular sieving. *Nat. Commun.* **2015**, *6*. [CrossRef] [PubMed]
50. Dendooven, J.; Ramachandran, R.K.; Devloo-Casier, K.; Rampelberg, G.; Filez, M.; Poelman, H.; Marin, G.B.; Fonda, E.; Detavernier, C. Low-Temperature Atomic Layer Deposition of Platinum Using (Methylcyclopentadienyl)trimethylplatinum and Ozone. *J. Phys. Chem. C* **2013**, *117*, 20557–20561. [CrossRef]

Article

Nickel Decorated on Phosphorous-Doped Carbon Nitride as an Efficient Photocatalyst for Reduction of Nitrobenzenes

Anurag Kumar [1,2], Pawan Kumar [1,2,†], Chetan Joshi [1,†], Manvi Manchanda [1], Rabah Boukherroub [3,*] and Suman L. Jain [1,*]

1 CSIR Indian Institute of Petroleum, Haridwar road Mohkampur, Dehradun 248005, India; anuragmnbd@gmail.com (A.K.); choudhary.2486pawan@yahoo.in (P.K.); chetanjoshi019@gmail.com (C.J.); manvimanchanda@gmail.com (M.M.)

2 Academy of Scientific and Industrial Research (AcSIR), New Delhi 110001, India

3 Institut d'Electronique, de Microélectronique et de Nanotechnologie (IEMN), UMR CNRS 8520, Université Lille1, Avenue Poincaré-BP 60069, 59652 Villeneuve d'Ascq, France

* Correspondence: rabah.boukherroub@iemn.univ-lille1.fr (R.B.); suman@iip.res.in (S.L.J.); Tel.: +91-135-2525-788 (S.L.J.); +33-0-3-62-53-17-24 (R.B.); Fax: +91-135-2660-098 (S.L.J.); +33-0-3-62-53-17-01 (R.B.)

† These authors contributed equally to this work.

Academic Editors: Hermenegildo García and Sergio Navalón
Received: 3 February 2016; Accepted: 21 March 2016; Published: 1 April 2016

Abstract: Nickel nanoparticle-decorated phosphorous-doped graphitic carbon nitride (Ni@g-PC_3N_4) was synthesized and used as an efficient photoactive catalyst for the reduction of various nitrobenzenes under visible light irradiation. Hydrazine monohydrate was used as the source of protons and electrons for the intended reaction. The developed photocatalyst was found to be highly active and afforded excellent product yields under mild experimental conditions. In addition, the photocatalyst could easily be recovered and reused for several runs without any detectable leaching during the reaction.

Keywords: carbon nitride; nickel nanoparticles; photocatalysis; visible light; nitrobenzene reduction

1. Introduction

Solar energy is abundant, inexpensive and has great potential for use as a clean and economical energy source for organic transformations [1–4]. Thus, the efficient conversion of solar energy to chemical energy, *i.e.*, visible light-driven chemical conversion (photocatalysis) is an area of tremendous importance and has been rapidly growing worldwide during the last two decades. The reduction of nitrobenzenes to the corresponding amines is an important transformation both in organic synthesis as well as in the chemical industry [5–7]. The conventional methods for this reaction mainly require harsh reaction conditions, expensive reagents, multi-step synthetic procedures and mostly generate hazardous waste [8,9]. Owing to the current need to develop greener and sustainable synthesis, the visible light-assisted reduction of nitrobenzenes to the corresponding amines is highly desired, as these reactions occur under mild and ambient conditions [10]. So far, semiconducting materials, mainly TiO_2-based heterogeneous photocatalysts, have been widely used for the reduction of nitrobenzenes [11]. However, these photocatalysts work only under ultraviolet (UV) irradiation, which is a small part of the solar spectrum and also needs special reaction vessels. In order to improve their efficiency in the visible region, surface modification of the TiO_2 photocatalyst by doping metal or metal oxides, oxide halides (*i.e.*, $PbPnO_2X$ (Pn = Bi, Sb; X = Br, Cl) and sensitization with dyes has also been demonstrated [12–14]. However, the tedious synthetic procedures and poor product yields limit the synthetic utility of these methods.

Another semiconducting material, *i.e.*, carbon nitride, has gained significant attention in recent years due to its low band gap (2.7 eV) and easy synthesis method [15–17]. The graphitic carbon nitride (g-C_3N_4) is a two dimensional (2D) polymer consisting of interconnected tri-s-triazine units via tertiary amines. The low band gap and the appropriate position of the conduction (−1.1 eV) and valence (+1.6 eV) bands enable it to initiate any redox reaction [18]. Despite of several advantages, it also suffers from certain drawbacks such as only being able to absorb blue light up to the 450 nm wavelength, so the red region of visible light cannot be harvested. In order to widen the absorption profile in the whole visible region, the doping of elements such as phosphorus (P), boron (B), fluorine (F), etc is important because these atoms form bonds with nitrogen and contribute their electrons to the π-conjugated system more efficiently. Recently, Zhang *et al.* synthesized P-doped g-C_3N_4 with an improved photocurrent response by using 1-butyl-3-methylimidazolium hexafluorophosphate ($BmimPF_6$) as a source of phosphorous [19]. Hong and coworkers synthesized sulfur-doped mesoporous carbon nitride (mpgCNS) using an *in situ* doping method, which showed increased photocatalytic performance for the solar hydrogen production [20]. Wang *et al.* synthesized fluorine-doped carbon nitride by using ammonium fluoride as a cheap source of fluorine for the enhanced visible light-driven hydrogen evolution reaction [21]. Very recently, our group reported an iron(II) bipyridine complex grafted nanoporous carbon nitride ($Fe(bpy)_3$/npg-C_3N_4) photocatalyst for the visible light-mediated oxidative coupling of benzylamines to imines [22].

In continuation of our ongoing research on the development of novel photocatalysts [23–25], herein we report for the first time on nickel nanoparticles grafted on P-doped g-C_3N_4 (Ni@g-PC_3N_4) as a high-performance photocatalyst for the reduction of nitrobenzenes to the corresponding amines at ambient temperature under visible light irradiation (Scheme 1). Due to better visible light absorption, P-doped carbon nitride can generate electron-hole pairs under visible light irradiation, while nickel nanoparticles, due to their lower Fermi level than conduction band of P-C_3N_4, can work as electron capturing agents to slow down the process of electron-hole recombination and enhance the catalytic activity and reaction rate.

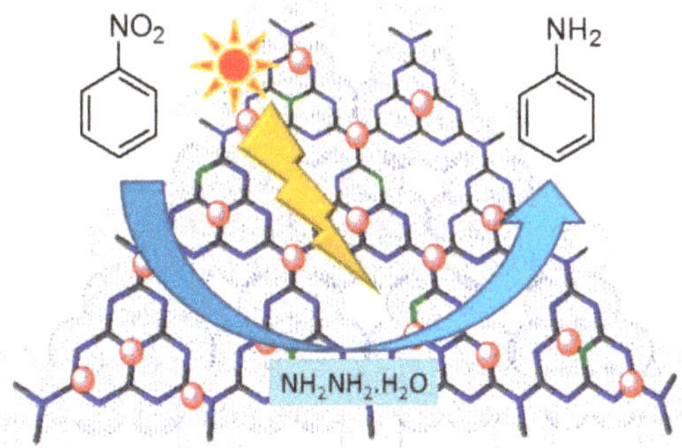

Scheme 1. Reduction of nitrobenzenes on Ni@g-PC_3N_4 (nickel nanoparticles grafted on P-doped g-C_3N_4) catalyst.

2. Results and Discussion

2.1. Synthesis and Characterization of the Photocatalyst

At first, nickel nanoparticles were synthesized by reducing nickel chloride hexahydrate by sodium borohydride in the presence of cetyltrimethylammonium bromide (CTAB) as a template by following the literature procedure in [26]. Next, the synthesis of phosphorous-doped graphitic carbon nitride (g-PC_3N_4) was performed by using ionic liquid 1-butyl-3-methylimidazolium hexafluorophosphate

[BmimPF$_6$] as a source of phosphorous and dicyandiamide as a precursor for the graphitic carbon nitride skeleton [17,27]. Nickel nanoparticle-decorated phosphorous-doped carbon nitride (Ni@g-PC$_3$N$_4$) with variable nickel content (2–7.5 wt %) was synthesized by the addition of nickel nanoparticles, BmimPF$_6$ and dicyandiamide into water with continuous stirring. The resulting suspension was heated at 100 °C until all the water was evaporated. Then, the obtained solid was heated at a programmed temperature in a similar manner to the one used for P-doped carbon nitride (Scheme 2). Bare graphitic carbon nitride (g-C$_3$N$_4$) was synthesized for comparison by heating dicyandiamide under identical programmed temperature. Among the various nanocomposites synthesized, 5% nickel nanoparticles grafted on P-doped g-C$_3$N$_4$ (5%Ni@PC$_3$N$_4$) exhibited the best performance for the photocatalytic reduction of nitrobenzene; thus, we have used this photocatalyst for detailed characterization and further studies.

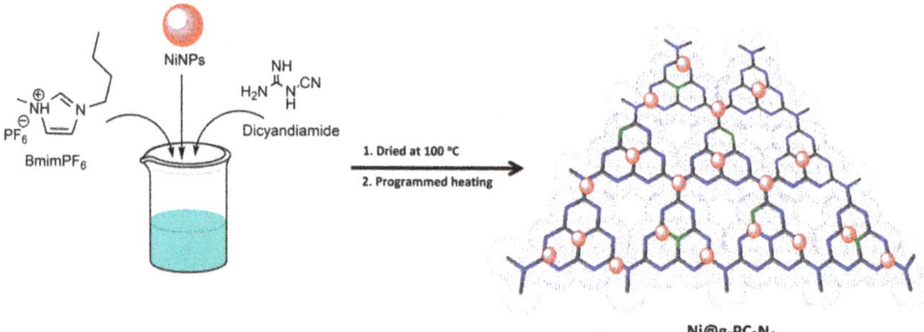

Scheme 2. Synthetic illustration of Ni@g-PC$_3$N$_4$ catalyst. NiNPs: nickel nanoparticles; BmimPF$_6$: 1-butyl-3-methylimidazolium hexafluorophosphate.

The surface morphology of the synthesized materials was determined by field emission scanning electron microscopy (FE-SEM). The graphitic carbon nitride (g-C$_3$N$_4$) showed many enfolded and crumpled sheet-like structures (Figure 1a). The framework of C$_3$N$_4$ contains nitrogen as a substituted heteroatom forming similarly to the π-conjugated system, as in graphitic planes, due to sp^2 hybridization between carbon and nitrogen atoms. The phosphorous-doped carbon nitride (g-PC$_3$N$_4$) exhibits similar morphological characteristics (Figure 1b). In the case of nickel nanoparticle-grafted g-PC$_3$N$_4$ (5%Ni@g-PC$_3$N$_4$), a similar sheet-like morphology was observed that may be due to the incorporation of nickel nanoparticles between the graphitic carbon nitride sheets (Figure 1c). The energy dispersive X-ray spectroscopy (EDX) pattern of g-PC$_3$N$_4$ clearly showed the presence of phosphorous (Figure 1e) and, in 5%Ni@g-PC$_3$N$_4$, the peak due to nickel can clearly be seen, which confirmed the presence of both components in the synthesized photocatalyst (Figure 1f).

The fine morphological features of the synthesized materials were determined with the help of high-resolution transmission electron microscopy (HRTEM). The transmission electron microscopy (TEM) image of g-C$_3$N$_4$ at 100 nm resolution displayed enfolded sheets of carbon nitride (Figure 2a). For g-PC$_3$N$_4$, similar folded sheet-like structures were observed, indicating that phosphorus doping did not change the structural features of the material (Figure 2b). In the TEM image of 5%Ni@g-PC$_3$N$_4$, several small dark spots were observed, suggesting the presence of nickel nanoparticles between the scaffolds of graphitic structure of phosphorous-doped carbon nitride (Figure 2c). The presence of several rings and bright spots in the selected area electron diffraction (SAED) pattern of 5%Ni@g-PC$_3$N$_4$ confirmed the presence of nickel nanoparticles in the composite structure (Figure 2d). Further EDX patterns of 5%Ni@g-PC$_3$N$_4$ revealed the presence of all the expected elements in the photocatalyst (Figure 2e).

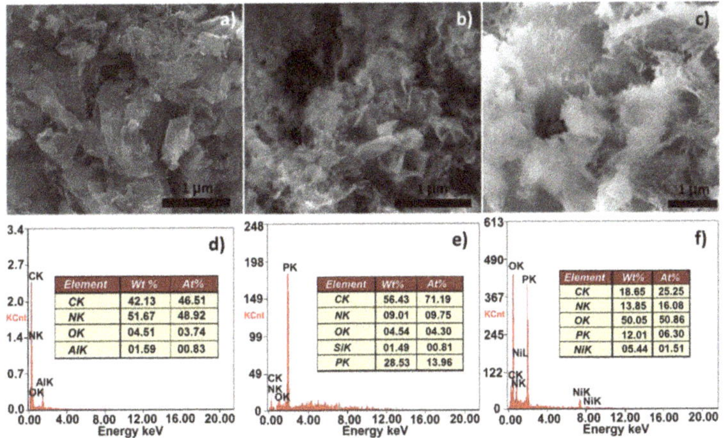

Figure 1. Field emission scanning electron microscopy (FE-SEM) image of: (**a**) graphitic carbon nitride (g-C_3N_4); (**b**) phosphorous-doped graphitic carbon nitride (g-PC_3N_4); (**c**) nickel nanoparticle-grafted g-PC_3N_4 (Ni@g-PC_3N_4); energy dispersive X-ray spectroscopy (EDX) pattern of: (**d**) g-C_3N_4; (**e**) g-PC_3N_4; (**f**) 5%Ni@g-PC_3N_4.

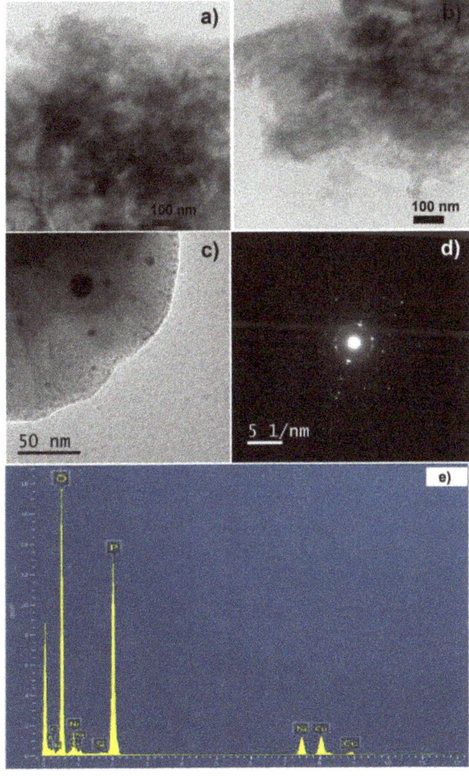

Figure 2. Transmission electron microscopy (TEM) images of: (**a**) g-C_3N_4; (**b**) g-PC_3N_4; (**c**) Ni@g-PC_3N_4; (**d**) selected area electron diffraction (SAED) pattern of Ni@g-PC_3N_4; (**e**) EDX pattern of Ni@g-PC_3N_4.

The vibrational spectra of the synthesized materials are depicted in Figure 3. The Fourier transform infrared (FTIR) spectrum of as-synthesized nickel nanoparticles by using cetyl trimethylammonium chloride as a template was found to be well in agreement with the known literature and gave peaks at 676 cm^{-1} due to the Ni–O stretch of oxidized nickel [28]. Some additional peaks were also identified that may be due to the residual template molecules in the particles. The FTIR spectrum of carbon nitride (g-C_3N_4) showed a characteristic peak at 815 cm^{-1} related to C–N heterocycles due to the triazine ring mode. Another peak in the range of 1200–1600 cm^{-1}, attributed to the specific aromatic skeleton vibration of carbon nitride, was observed [29]. A broad band was evident in the range of 3000–3700 cm^{-1} corresponding to adsorbed moisture, *i.e.*, the presence of water molecules, or due to the stretching mode of –NH_2, which are uncondensed amine groups, or to N–H group vibrations present at the surface of carbon nitride (Figure 3a). For the P-doped carbon nitride (g-PC_3N_4), peaks related to C–N at 1520, 1440 and 1310 cm^{-1} were found to be diminished, which is mainly due to the displacement of some carbon atoms in the ring skeleton by the phosphorous atoms to give P–N bonds. Furthermore, a new peak at 980 cm^{-1} due to the vibration of the C–P heterocycle was clearly observed (Figure 3c). For the 5%Ni@g-PC_3N_4, the appearance of some peaks due to nickel nanoparticles with g-PC_3N_4 confirmed the successful synthesis of nickel-grafted P-doped g-C_3N_4 (Figure 3d).

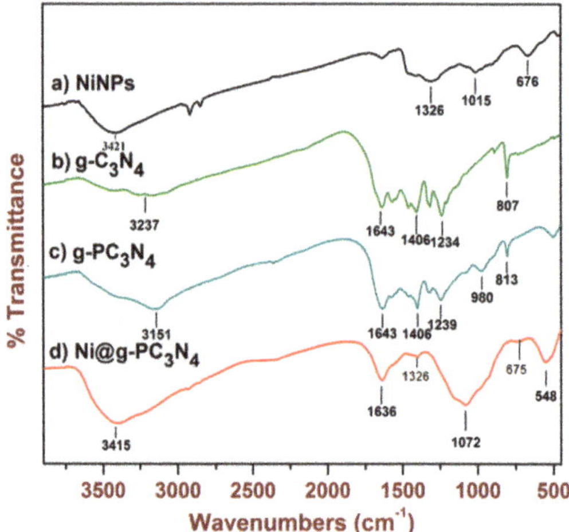

Figure 3. Fourier transform infrared (FTIR) spectra of: (**a**) nickel nanoparticles (NiNPs); (**b**) g-C_3N_4; (**c**) g-PC_3N_4; (**d**) 5%Ni@g-PC_3N_4.

The crystallinity and phase structure of the nanocomposites were determined by X-ray diffraction (XRD) (Figure 4). The XRD diffraction pattern of graphitic carbon nitride (g-C_3N_4) revealed an intense broad peak at the 2θ value of 27.4°, indexed to (002) planes with 0.32 nm interlayer distance due to the stacking of graphite-like conjugated triazine aromatic sheets, which matches well with Joint committee on powder diffraction standards (JCPDS) 87–1526 for g-C_3N_4 (Figure 4a) [30]. The g-PC_3N_4 also exhibited an identical diffraction pattern; however, the intensity of the peak at 27.4° was found to be reduced due to the replacement of carbon atoms by phosphorous atoms (Figure 4b). The diffraction pattern of the 5%Ni@g-PC_3N_4 did not show any peak of nickel nanoparticles. This is most likely due to the amorphous nature of the material (Figure 4c). Furthermore, the peak at 27.4° was absent in 5%Ni@g-PC_3N_4, which was most probably due to the intercalation of nickel nanoparticles between the sheets. Therefore, the interlayer distance between sheets increased and the diffraction due to stacked sheets disappeared.

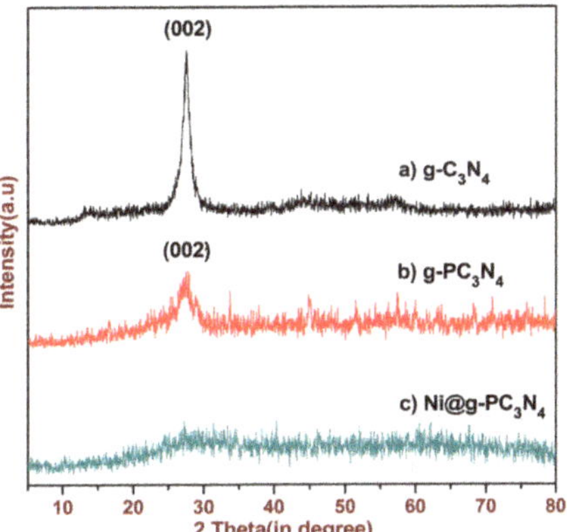

Figure 4. X-ray diffraction (XRD) patterns of: (**a**) g-C_3N_4; (**b**) g-PC_3N_4; (**c**) 5%Ni@g-PC_3N_4. a.u.: arbitrary units.

To determine the surface properties of the synthesized materials, N_2 adsorption-desorption isotherms were determined by using multilayer Brunauer–Emmett–Teller (BET) adsorption-desorption theory. As seen in Figure 5, the adsorption-desorption isotherms were of type IV, suggesting the mesoporous nature of the materials [31]. The BET surface area (S_{BET}), total pore volume (V_p) and mean pore diameter (r_p) of g-C_3N_4 were found to be 14.67 $m^2 \cdot g^{-1}$, 0.15 $cm^3 \cdot g^{-1}$ and 4.21 nm, while for g-PC_3N_4 these values were determined to be 24.59 $m^2 \cdot g^{-1}$, 0.14 $cm^3 \cdot g^{-1}$ and 3.83 nm, respectively. For the 5%Ni@g-PC_3N_4 surface area (S_{BET}), the total pore volume (V_p) and mean pore diameter (r_p) were 48.62 $m^2 \cdot g^{-1}$, 0.17 $cm^3 \cdot g^{-1}$ and 2.64 nm, respectively. The change in the surface properties clearly indicated the grafting of nickel nanoparticles between the sheets of carbon nitride (Figure 5). The higher surface area provides more sites for reaction to proceed on the surface of the Ni@g-PC_3N_4 catalyst.

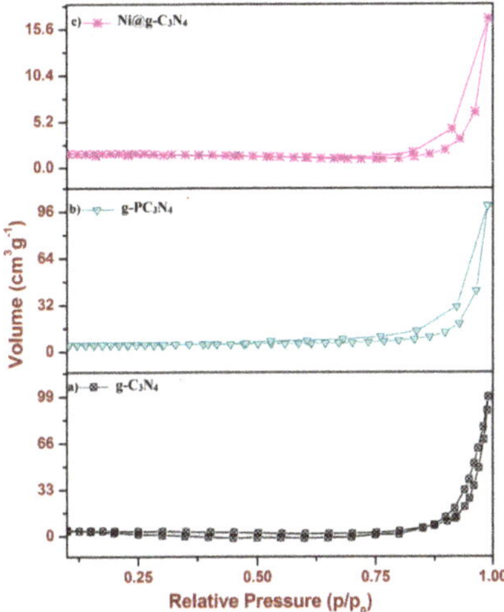

Figure 5. N_2 adsorption-desorption isotherm and pore size distribution of: (**a**) g-C_3N_4; (**b**) g-PC_3N_4; (**c**) 5%Ni@g-PC_3N_4.

Electronic and optical properties are of fundamental importance to determine the photo activity of a catalyst. The absorption properties of synthesized materials were determined by UV-visible

(UV-Vis) spectroscopy. The UV-visible spectrum of pure g-C_3N_4 shows an absorption spectrum similar to a typical semiconductor absorption spectrum between 200–450 nm, originating from the charge transfer from a populated valence band of nitrogen atom (2p orbitals) to a conduction band of carbon atom (2p orbitals) of carbon nitride [32]. An additional sharp peak at 256 nm, attributed to the aromatic ring's $\pi \rightarrow \pi^*$ transition, was observed. Another intense peak at 384 nm, due to the $n \rightarrow \pi^*$ transitions caused by the electron transfer from a nitrogen nonbonding orbital to an aromatic anti-bonding orbital, was also present in the UV-Vis spectrum. The band tailing from 410 to 500 nm suggests a slight visible light absorption capacity of carbon nitride (Figure 6a). After doping with phosphorous atoms, the absorption band due to the nitrogen nonbonding to aromatic antibonding ($n \rightarrow \pi^*$) transition was diminished due to the replacement of carbon atoms with phosphorous, whereas the absorption pattern was found to be redshifted due to the better charge contribution by loosely bound electrons of phosphorous in the aromatic conjugated system (Figure 6b) [33]. After incorporation of nickel nanoparticles (5%Ni@g-PC_3N_4), the material exhibited lower absorbance in the visible region as the nickel nanoparticles have lower absorbance in the visible region, and therefore reduce the absorption profile in the visible region (Figure 6c).

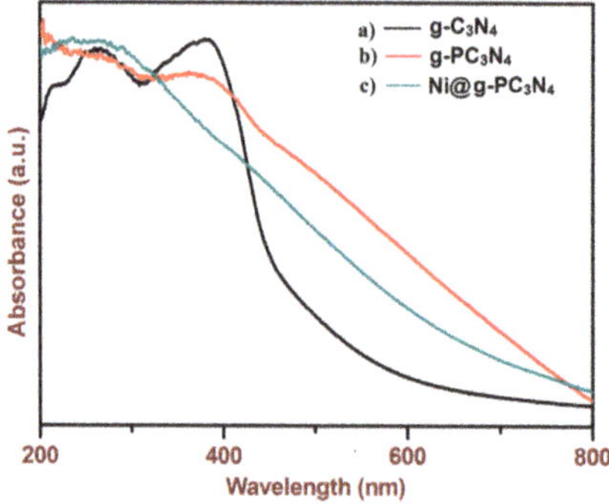

Figure 6. Ultraviolet–visible (UV-Vis) absorption spectra of: (**a**) g-C_3N_4; (**b**) g-PC_3N_4; (**c**) 5%Ni@g-PC_3N_4.

In order to confirm the visible light absorption of the synthesized photocatalysts, a Tauc plot was obtained as shown in Figure 7. From the Tauc plot, the band gap value for the g-C_3N_4 was determined to be 2.69 eV due to the uneven synthesis of sheets, and was well in concordance with the existing literature [18]. After doping with phosphorous atoms (g-PC_3N_4), the value of the band gap decreased to 1.45 eV, which may be due to the displacement of carbon atoms by the phosphorous atoms in the triazine ring skeletons of carbon nitride. For the 5%Ni@g-PC_3N_4, a band gap value of 1.52 eV was calculated, which is nearly similar to that of g-PC_3N_4.

The thermal stability of the synthesized materials was examined by thermogravimetric analysis (Figure 8). A thermogravimetric analysis (TGA) pattern of pristine nickel nanoparticles showed steady weight loss (approx. 15%) from 100 to 500 °C, which may be due to loss of the residual organic template (CTAB) and loss of other oxygen-carrying functionalities during the crystallization step (Figure 8a). The thermogram of g-C_3N_4 exhibited a sharp weight loss in the range of 560 to 720 °C (Figure 8b). The P-doped g-C_3N_4 displayed a similar weight loss pattern as g-C_3N_4 (Figure 8c). For the 5%Ni@g-PC_3N_4, a small weight loss at around 100 °C was observed due to the loss of moisture, followed by steady weight loss up to 700 °C due to the loss of template in nickel and the slow degradation of the phosphorous-doped sheets. At 770 °C, a sharp weight loss was observed due to the complete degradation of P-doped carbon nitride sheets (Figure 8d).

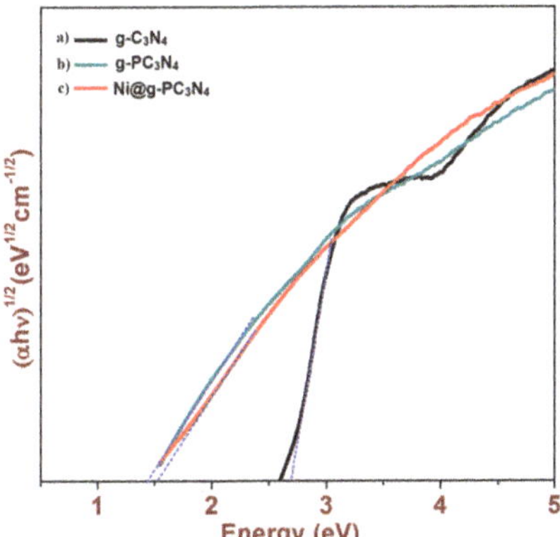

Figure 7. Tauc plots of (**a**) g-C_3N_4; (**b**) g-PC_3N_4; (**c**) Ni@g-PC_3N_4. α: absoption coefficient; hν: energy of incident photon.

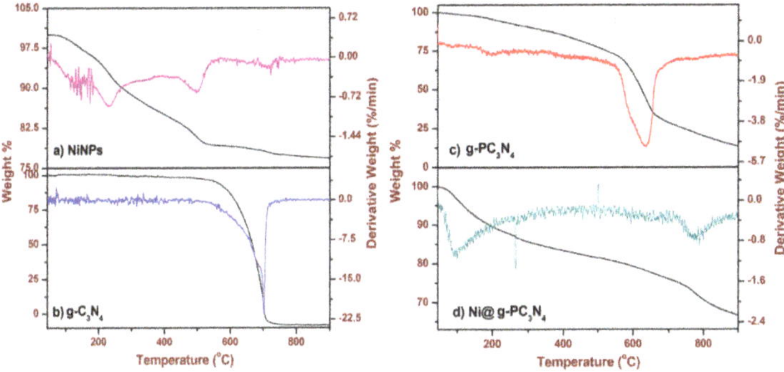

Figure 8. Thermogravimetric analysis (TGA) spectra of: (**a**) NiNPs; (**b**) g-C_3N_4; (**c**) g-PC_3N_4; (**d**) 5%Ni@g-PC_3N_4.

2.2. Photocatalytic Reduction Reaction

The photocatalytic activity of the synthesized NiNPs, g-C_3N_4, g-PC_3N_4 and 2–7.5%Ni@g-PC_3N_4 catalysts was tested for the reduction of nitrobenzene as a model substrate using hydrazine monohydrate as a proton source under visible light irradiation. The results of these optimization experiments are summarized in Table 1. There was no conversion obtained with pristine nickel nanoparticles (NiNPs) (Table 1, entry 1). However, the yield of product was found to be 24.6 % when g-C_3N_4 was used as a photocatalyst (Table 1, entry 2). In case of P-doped g-PC_3N_4, the yield of aniline increased up to 54.2%, which clearly indicated the promoting effect of P-doping due to the better visible light absorption (Table 1, entry 3). Furthermore, the yield of aniline increased to manifold by using Ni@g-PC_3N_4 and, after 8 h of visible light irradiation, the yield reached 96.5% (Table 1, entries 4–6). This increase in the reaction rate and product yield can be explained because the nickel nanoparticles work as electron sinks, and the photogenerated electrons flow from conduction band of g-PC_3N_4 to nickel nanoparticles. This makes the electron unavailable for the recombination and therefore increases the activity of the catalyst. We also determined the optimum nickel content in the photocatalyst. It was found that among all the synthesized Ni@g-PC_3N_4 nanocomposites having variable nickel contents (2–7.5 wt % Ni), the nanocomposite having 5 wt % nickel nanoparticles was the most active and afforded maximum product yield (Table 1, entry 5). In the case of 2%Ni, the yield of product was lower, whereas no significant enhancement was observed with increasing the concentration of Ni beyond 5% (Table 1, entry 6). Furthermore, the reaction did not take place in dark conditions at ambient temperature by using g-C_3N_4, g-PC_3N_4 and Ni@g-PC_3N_4 photocatalysts, which clearly revealed that the reaction was visible light promoted (Table 1). The use of hydrazine hydrate was found to be essential as it provided required protons and no product was formed in the absence of hydrazine hydrate (Table 1, entry 5).

Table 1. Results of optimization experiments [a]. NiNPs: nickel nanoparticles; g-C_3N_4: graphitic carbon nitride; g-PC_3N_4: phosphorous doped g-C_3N_4; Ni@g-PC_3N_4: nickel nanoparticles grafted on P-doped g-C_3N_4; TOF: turn over frequency.

Entry	Catalyst	Conditions	Time (h)	AnilineYield (%) [b]	TOF (h^{-1})
1	NiNPs	Dark	24	-	-
		Visible light	-	-	-
2	g-C_3N_4	Dark	24	-	-
		Visible light	12	24.6	2.0
3	g-P-C_3N_4	Dark	24	-	-
		Visible light	12	54.2	4.5
4	2%Ni@g-PC_3N_4	Dark	24	-	-
		Visible light	8	82.0	10.2
5	5%Ni@g-PC_3N_4	Dark	24	Trace	-
		Visible light	8	96.5	12.1
		Visible light	24 [c]	- [c]	- [c]
6	7.5%Ni@g-PC_3N_4	Dark	24	10	0.4
		Visible light	8	97.0	12.1

[a] Reaction conditions: nitrobenzene, 0.1 mmol; catalyst, 25 mg; hydrazine monohydrate, 1 mmol; Irradiation, White cold 20 W light emitting diode (LED) $\lambda > 400$ nm, Power at reaction vessel 70 W/m^2; [b] Isolated yield; [c] without hydrazine monohydrate.

Based on these optimization experiments, 5%Ni@g-PC_3N_4 was selected as the optimum catalyst for further studies. The reaction was further generalized for various substituted nitrobenzenes and the results are summarized in Table 2. It can be seen that the substituent effect did not play much of a role and excellent product yields were obtained in all cases within 8 to 10.5 h.

Table 2. 5%Ni@g-PC_3N_4 catalyzed photoreduction of nitrobenzenes [a].

Entry	Reactant	Product	Time/h	Conversion (%) [b]	Yield (%) [c]	TOF (h^{-1})
1	$C_6H_5{-}NO_2$	$C_6H_5{-}NH_2$	8.0	98.0	96.5	12.1
2	$HO{-}C_6H_4{-}NO_2$	$HO{-}C_6H_4{-}NH_2$	8.0	95.5	94.2	11.7
3	$H_3C{-}C_6H_4{-}NO_2$	$H_3C{-}C_6H_4{-}NH_2$	8.0	96.4	95.4	11.9
4	$(CH_3)_3C{-}C_6H_4{-}NO_2$	$(CH_3)_3C{-}C_6H_4{-}NH_2$	8.0	94.8	93.0	11.6
5	$H_3CO{-}C_6H_4{-}NO_2$	$H_3CO{-}C_6H_4{-}NH_2$	8.0	95.6	94.2	11.7
6	$C_2H_5O{-}C_6H_4{-}NO_2$	$C_2H_5O{-}C_6H_4{-}NH_2$	8.0	96.0	94.6	11.8
7	$Cl{-}C_6H_4{-}NO_2$	$Cl{-}C_6H_4{-}NH_2$	8.5	90.5	89.4	10.5
8	$Br{-}C_6H_4{-}NO_2$	$Br{-}C_6H_4{-}NH_2$	8.5	92.4	91.0	10.7
9	$I{-}C_6H_4{-}NO_2$	$I{-}C_6H_4{-}NH_2$	10.5	93.6	92.8	8.8
10	$F_3C{-}C_6H_4{-}NO_2$	$F_3C{-}C_6H_4{-}NH_2$	10.5	90.4	89.6	8.5

[a] Reaction conditions: nitro compound, 0.1 mmol; catalyst, 25 mg; hydrazine monohydrate, 1 mmol; Irradiation, White cold 20 W LED $\lambda > 400$ nm, Power at reaction vessel 70 W/m^2; [b] Conversion (Conv.) was determined with gas chromatography (GC); [c] Isolated yield.

The experiments were performed to probe the heterogeneous nature and reusability of the photocatalyst. After the completion of the reaction, the photocatalyst was recovered by centrifugation, washed with methanol and dried at 50 °C. The recovered photocatalyst was reused for five subsequent runs under described experimental conditions. No significant loss was observed in the activity of the recycled catalyst, and the product yield remained almost unchanged even after five recycling experiments, which confirmed the true heterogeneous nature of the developed photocatalyst. Furthermore, an inductively coupled plasma-atomic emission spectrometry (ICP-AES) analysis of the photocatalyst after five recycling experiments showed a nickel content of 4.92 wt %, which is nearly similar to that of a fresh catalyst (4.98 wt %). These results confirmed that the developed photocatalyst was truly heterogeneous in nature and had not shown any detectable leaching during the reaction (Figure 9).

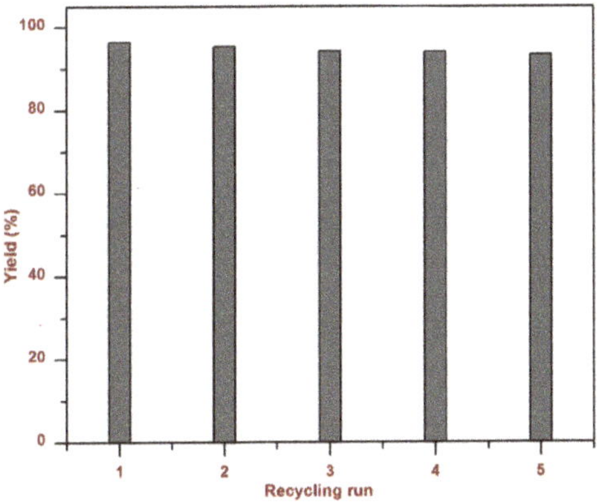

Figure 9. Results of recycling experiments.

Although the exact mechanism of the reaction is not clear at this stage, we assume that the generation of electron-hole pairs after the absorption of visible light is the first step to initiate the reaction [34]. Since the optical band gap of P-doped g-PC_3N_4 is 1.52 eV, it can generate electrons and holes in its conduction and valence band, respectively. Electrons from the conduction band of P-doped carbon nitride get transferred to nickel nanoparticles due to lower Fermi level, which therefore work as electron sinks and capture the photogenerated electrons and prevent back charge recombination [35]. The electrons from NiNPs are transferred to nitrobenzene, which initiate a one electron reduction step [36]. The holes in the valence band of g-PC_3N_4 oxidize hydrazine hydrate and generate electrons and protons along with nitrogen gas. The protons were used for the hydrogenation of activated molecules of nitrobenzene. As for the reduction of nitrobenezene to aniline, six protons and six electrons are required, so 3/2 $NH_2NH_2 \cdot H_2O$ mole were consumed per mole of reactant (Scheme 3).

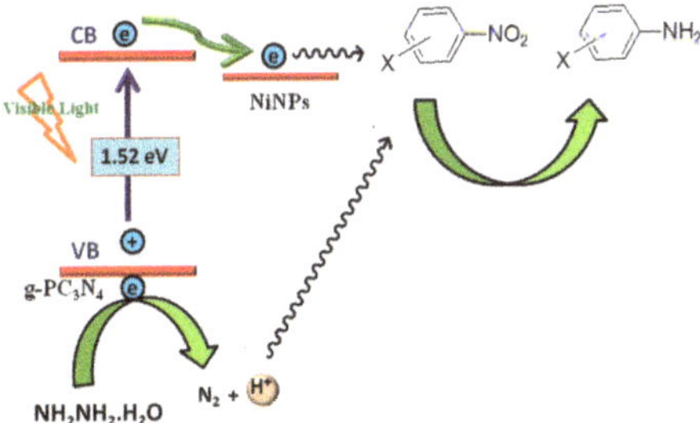

Scheme 3. Plausible mechanism on the basis of the band gap structure for the visible light reduction of nitrobenzenes by Ni@g-PC_3N_4 photocatalyst. CB: conduction band; VB: valence band.

3. Experimental Section

3.1. Materials

Dicyandiamide (99%), 1-butyl-3-methylimidazolium hexafluorophosphate [$BmimPF_6$] (⩾98.5%), nickel chloride hexahydrate (⩾98%), cetyltrimethylammonium bromide (⩾98%) and sodium borohydride (⩾98.0%) were purchased from Aldrich (St. Louis, MO, USA) and used as received. All substrates and solvents were purchased from Merck India Ltd (Mumbai, Maharashtra, India) and used without further purification.

3.2. Characterizations

The rough morphological surface properties of materials were determined with the help of scanning electron microscopy (SEM) using FE-SEM (Jeol Model JSM-6340F (Tokyo, Japan). For elaborating the fine structure of the catalysts, high-resolution transmission electron microscopy (HRTEM) was used and HRTEM images were collected on FEI-TecnaiG^2 Twin TEM (Hillsboro, OR, USA) operating at an acceleration voltage of 200 kV. For sample preparation, a very dilute suspension of material was deposited on a carbon coated TEM grid. TEM images were processed on GATAN micrograph software (Munchen, Germany). The vibration spectra (FT-IR) of samples were recorded on a Perkin–Elmer spectrum RX-1 IR spectrophotometer (Waltham, MA, USA) using a potassium bromide window. X-ray diffraction patterns for determining the phase structure and crystalline properties of the materials were obtained on a Bruker D8 Advance diffractometer (Billerica, MA, USA) working at 40 kV and 40 mA with Cu K_α radiation (λ = 0.15418 nm). Nitrogen adsorption-desorption isotherms for calculating surface properties like Brunauer-Emmet-Teller surface area (S_{BET}), Barret-Joiner-Halenda (BJH) porosity, pore volume, *etc.*, were acquired on a V_P. Micromeritics ASAP2010 (Norcross, GA, USA) at 77 K. UV-Vis absorption spectra of solid samples were collected on a Perkin Elmer lambda-19 UV-VIS-NIR spectrophotometer (Waltham, MA, USA) by making 1 mm thick pellets using $BaSO_4$ as reference. To check the thermal stability of materials, thermogravimetric analysis (TGA) was performed by using a TA-SDT Q-600 thermal analyser (Champaign, IL, USA) in the temperature range of 45 to 800 °C under nitrogen flow with a heating rate of 10 °C/min. To determine the nickel content in the synthesized composites, ICP-AES analysis was performed by using inductively coupled plasma atomic emission spectrometer (ICP-AES, DRE, PS-3000UV, Leeman Labs Inc., Hudson, NH, USA). Samples for ICP-AES were made by digesting a calculated amount of samples with nitric acid followed by filtration and making volume up to 10 mL by adding deionized water.

3.3. Synthesis of Nickel Nanoparticles [26]

Nickel nanoparticles were synthesized by following a literature procedure. In brief, to an aqueous suspension of nickel chloride hexahydrate (0.152 mmol, 0.019 g) and cetyltrimethyl ammonium bromide, CTAB (0.288 mmol, 0.105 g), an aqueous solution (1.5 mL) of $NaBH_4$ (0.020 g, 0.526 mmol) was added dropwise. The mixture was vigorously stirred to obtain a black suspension. The particles were separated by centrifugation and washed with water several times.

3.4. Synthesis of Phosphorous-Doped Graphitic Carbon Nitride (g-PC_3N_4) [19,27]

Phosphorous-doped graphitic carbon nitride was synthesized by using $BmimPF_4$ as a source of phosphorous. The phosphorous containing ionic liquid, $BmimPF_4$ (0.5 g) was dissolved in water (6 mL) and stirred for 5 min. After that, dicyandiamide (1 g) was added to this solution and the mixture was heated at 100 °C until all the water had completely evaporated, which resulted in the formation of a white solid. The obtained solid was subjected to heating at a programmed temperature. Firstly the sample was heated up to 350 °C within 2 h, then the temperature was maintained at a constant for the next 4 h. The temperature was then raised to 550 °C in 1 h and then this temperature was maintained at a constant for the next 4 h. The sample was collected at room temperature.

3.5. *Synthesis of Nickel Nanoparticles Decorated on Phosphorous-Doped Graphitic Carbon Nitride (Ni@g-PC_3N_4)*

For the synthesis of nickel nanoparticle-decorated carbon nitride, nickel particles were added along with $BmimPF_4$ and dicyandiamide during the synthesis step. Then the sample was heated at a programmed temperature as in the synthesis of P-doped carbon nitride.

3.6. *Photocatalytic Reduction Experiment*

To check the activity of the catalyst for the visible light mediated reduction, a 20 watt LED (Model No. HP-FL-20W-F-Hope LED Opto-Electric Co., Ltd (Shenzhen, China), $\lambda > 400$ nm) was used as a source of visible light. A round bottom flask was charged with 25 mg of 5%Ni@g-PC_3N_4 catalyst, 0.1 mmol of nitrobenzene and 1.0 mmol of hydrazine monohydrate in 10 mL solvent mixture of acetonitrile/DCM/methanol. After sonication for 10 min, the resulting mixture was irradiated under visible light. To monitor the progress of the reaction, the sample was withdrawn at a certain period of time and analyzed by gas chromatography-flame ionization detector (GC-FID). After the completion of the reaction, the solvent was removed by rotary evaporation and the product was isolated using column chromatography. The identification of product was done by gas chromatography-mass spectrometry (GC-MS) and ^{1}H-nuclear magnetic resonance (^{1}H-NMR).

4. Conclusions

We have synthesized a novel, highly efficient and visible light-active nickel nanoparticles-decorated P-doped carbon nitride for the selective hydrogenation of nitro compounds to the corresponding amines in the presence of hydrazine monohydrate as a proton donor. Due to P-doping, the developed photocatalyst absorbs the maximum part of the visible region and nickel nanoparticles work as an electron trap. The developed photocatalyst was found to be highly effective and afforded excellent product yields within 8–10.5 h at ambient temperature under visible light irradiation. Due to the heterogeneous nature of photocatalyst, it can be easily recovered and reused for further reactions without any significant decrease in its activity. The enhanced performance of the catalyst in the visible region can be explained on the basis of reduced band gap, which generates electron-hole pairs through the absorption of visible light. Photogenerated electrons are efficiently captured by nickel nanoparticles on the sheets and subsequently transferred to substrate molecule, while holes are used to oxidize hydrazine hydrate and extract required protons and electrons for the reaction.

Acknowledgments: Authors are thankful to Director IIP for granting permission to publish these results. Anurag Kumar and Pawan Kumar are thankful to Council of Scientific and Industrial Research (CSIR) New Delhi for providing research fellowships. Chetan Joshi kindly acknowledges CSIR, New Delhi for providing technical Human Resource (H.R.) under XII five year projects. Manvi Manchanda acknowledges Department of Science and Technology (DST) New Delhi, for providing fellowship under the Women Societal B (WoS-B) program. Analytical department is kindly acknowledged for the analysis of samples. Rabah Boukherroub acknowledges financial support from the Centre National de la Recherche Scientifique (CNRS), Lille1 University and Nord Pas de Calais region.

Author Contributions: Anurag Kumar, Pawan Kumar, Chetan Joshi and Manvi Manchanda were involved in the synthesis, characterization and catalytic evaluation of photocatalyst. Suman L. Jain and Rabah Boukherroub helped in the technical discussion, interpretation of data and writing of the manuscript.

Conflicts of Interest: The authors declare no conflict of interest.

References

1. Narayanam, J.M.R.; Stephenson, C.R.J. Visible light photoredox catalysis: Applications in organic synthesis. *Chem. Soc. Rev.* **2011**, *40*, 102–113. [CrossRef] [PubMed]
2. Yoon, T.P.; Ischay, M.A.; Du, J. Visible light photocatalysis as a greener approach to photochemical synthesis. *Nat. Chem.* **2010**, *2*, 527–532. [CrossRef] [PubMed]
3. Lang, X.; Chen, X.; Zhao, J. Heterogeneous visible light photocatalysis for selective organic transformations. *Chem. Soc. Rev.* **2014**, *43*, 473–486. [CrossRef] [PubMed]

4. Su, F.; Mathew, S.C.; Lipner, G.; Fu, X.; Antonietti, M.; Blechert, S.; Wang, X. mpg-C_3N_4-Catalyzed Selective Oxidation of Alcohols Using O_2 and Visible Light. *J. Am. Chem. Soc.* **2010**, *132*, 16299–16301. [CrossRef] [PubMed]
5. Wang, A.J.; Cheng, H.Y.; Liang, B.; Ren, N.Q.; Cui, D.; Lin, N.; Kim, B.H.; Rabaey, K. Efficient reduction of nitrobenzene to aniline with a biocatalyzed cathode. *Environ. Sci. Technol.* **2011**, *45*, 10186–10193. [CrossRef] [PubMed]
6. Corma, A.; Concepcion, P.; Serna, P. A different reaction pathway for the reduction of aromatic nitro compounds on gold catalysts. *Angew. Chem. Int. Ed.* **2007**, *46*, 7266–7269. [CrossRef] [PubMed]
7. Wang, J.; Yuan, Z.; Nie, R.; Hou, Z.; Zheng, X. Hydrogenation of nitrobenzene to aniline over silica gel supported nickel catalysts. *Ind. Eng. Chem. Res.* **2010**, *49*, 4664–4669. [CrossRef]
8. Kulkarni, A.S.; Jayaram, R.V. Liquid phase catalytic transfer hydrogenation of aromatic nitro compounds on perovskites prepared by microwave irradiation. *Appl. Catal. A* **2013**, *252*, 225–230. [CrossRef]
9. Xu, W.Y.; Gao, T.Y.; Fan, J.H. Reduction of nitrobenzene by the catalyzed Fe–Cu process. *J. Hazard. Mater.* **2005**, *123*, 232–241. [CrossRef] [PubMed]
10. Agrawal, A.; Tratnyek, P.G. Reduction of Nitro Aromatic Compounds by Zero-Valent Iron Metal. *Environ. Sci. Technol.* **1995**, *30*, 153–160. [CrossRef]
11. Tanaka, A.; Nishino, Y.; Sakaguchi, S.; Yoshikawa, T.; Imamura, K.; Hashimoto, K.; Kominami, H. Functionalization of a plasmonic Au/TiO_2 photocatalyst with an Ag co-catalyst for quantitative reduction of nitrobenzene to aniline in 2-propanol suspensions under irradiation of visible light. *Chem. Commun.* **2013**, *49*, 2551–2553. [CrossRef] [PubMed]
12. Füldner, S.; Pohla, P.; Bartling, H.; Dankesreiter, S.; Stadler, R.; Gruber, M.; Pfitzner, A.; König, B. Selective photocatalytic reductions of nitrobenzene derivatives using $PbBiO_2X$ and blue light. *Green Chem.* **2011**, *13*, 640–643. [CrossRef]
13. Richner, G.; Bokhoven, J.A.; van Neuhold, Y.M.; Makosch, M.; Hungerbühler, K. *In situ* infrared monitoring of the solid/liquid catalyst interface during the three-phase hydrogenation of nitrobenzene over nanosized Au on TiO_2. *Phys. Chem. Chem. Phys.* **2011**, *13*, 12463–12471. [CrossRef] [PubMed]
14. Huang, H.; Zhou, J.; Liu, H.; Zhou, Y.; Feng, Y. Selective photoreduction of nitrobenzene to aniline on TiO_2 nanoparticles modified with amino acid. *J. Hazard. Mater.* **2010**, *178*, 994–998. [CrossRef] [PubMed]
15. Wang, X.; Maeda, K.; Chen, X.; Takanabe, K.; Domen, K.; Hou, Y.; Fu, X.; Antonietti, M. Polymer semiconductors for artificial photosynthesis: Hydrogen evolution by mesoporous graphitic carbon nitride with visible light. *J. Am. Chem. Soc.* **2009**, *131*, 1680–1681. [CrossRef] [PubMed]
16. Wang, X.; Blechert, S.; Antonietti, M. Polymeric graphitic carbon nitride for heterogeneous photocatalysis. *ACS Catal.* **2012**, *2*, 1596–1606. [CrossRef]
17. Cui, Y.; Huang, J.; Fu, X.; Wang, X. Metal-free photocatalytic degradation of 4-chlorophenol in water by mesoporous carbon nitride semiconductors. *Catal. Sci. Technol.* **2012**, *2*, 1396–1402. [CrossRef]
18. Yan, S.C.; Li, Z.S.; Zou, Z.G. Photodegradation performance of g-C_3N_4 fabricated by directly heating melamine. *Langmuir* **2009**, *25*, 10397–10401. [CrossRef] [PubMed]
19. Zhang, Y.; Mori, T.; Ye, J.; Antonietti, M. Phosphorus-doped carbon nitride solid: enhanced electrical conductivity and photocurrent generation. *J. Am. Chem. Soc.* **2010**, *132*, 6294–6295. [CrossRef] [PubMed]
20. Hong, J.; Xia, X.; Wang, Y.; Xu, R. Mesoporous carbon nitride with *in situ* sulfur doping for enhanced photocatalytic hydrogen evolution from water under visible light. *J. Mater. Chem.* **2012**, *22*. [CrossRef]
21. Wang, Y.; Di, Y.; Antonietti, M.; Li, H.; Chen, X.; Wang, X. Excellent visible-light photocatalysis of fluorinated polymeric carbon nitride solids. *Chem. Mater.* **2010**, *22*, 5119–5121. [CrossRef]
22. Kumar, A.; Kumar, P.; Joshi, C.; Ponnada, S.; Pathak, A.K.; Ali, A.; Sreedhar, B.; Jain, S.L. A $[Fe(bpy)_3]^{2+}$ grafted graphitic carbon nitride hybrid for visible light assisted oxidative coupling of benzylamines under mild reaction conditions. *Green Chem.* **2016**. [CrossRef]
23. Kumar, S.; Kumar, P.; Deb, A.; Maiti, D.; Jain, S.L. Graphene oxide grafted with iridium complex as a superior heterogeneous catalyst for chemical fixation of carbon dioxide to dimethylformamide. *Carbon* **2016**, *100*, 632–640. [CrossRef]
24. Kumar, P.; Bansiwal, A.; Labhsetwar, N.; Jain, S.L. Visible light assisted photocatalytic reduction of CO_2 using a graphene oxide supported heteroleptic ruthenium complex. *Green Chem.* **2015**, *17*, 1605–1609. [CrossRef]

25. Gusain, R.; Kumar, P.; Sharma, O.P.; Jain, S.L.; Khatri, O.P. Reduced graphene oxide–CuO nanocomposites for photocatalytic conversion of CO_2 into methanol under visible light irradiation. *Appl. Catal. B* **2016**, *181*, 352–362. [CrossRef]
26. Singh, S.K.; Xu, Q. Bimetallic Ni-Pt Nanocatalysts for selective decomposition of hydrazine in aqueous solution to hydrogen at room temperature for chemical hydrogen storage. *Inorg. Chem.* **2010**, *49*, 6148–6152. [CrossRef] [PubMed]
27. Wang, Y.; Zhang, J.; Wang, X.; Antonietti, M.; Li, H. Boron-and fluorine-containing mesoporous carbon nitride polymers: Metal-free catalysts for cyclohexane oxidation. *Angew. Chem. Int. Ed.* **2010**, *49*, 3356–3359. [CrossRef] [PubMed]
28. Chen, D.H.; Wu, S.H. Synthesis of nickel nanoparticles in water-in-oil microemulsions. *Chem. Mater.* **2000**, *12*, 1354–1360. [CrossRef]
29. Liu, J.; Zhang, T.; Wang, Z.; Dawson, G.; Chen, W. Simple pyrolysis of urea into graphitic carbon nitride with recyclable adsorption and photocatalytic activity. *J. Mater. Chem.* **2011**, *21*, 14398–14401. [CrossRef]
30. Ge, L. Synthesis and photocatalytic performance of novel metal-free g-C_3N_4 photocatalysts. *Mater. Lett.* **2011**, *65*, 2652–2654. [CrossRef]
31. Rouquerol, J.; Avnir, D.; Fairbridge, C.W.; Everett, D.H.; Haynes, J.M.; Pernicone, N.; Ramsay, J.D.F.; Sing, K.S.W.; Unger, K.K. Recommendations for the characterization of porous solids (Technical Report). *Pure Appl. Chem.* **1994**, *66*, 1739–1758. [CrossRef]
32. Wang, Y.; Wang, X.; Antonietti, M. Polymeric graphitic carbon nitride as a heterogeneous organocatalyst: from photochemistry to multipurpose catalysis to sustainable chemistry. *Angew. Chem. Int. Ed.* **2012**, *51*, 68–89. [CrossRef] [PubMed]
33. Su, J.; Geng, P.; Li, X.; Zhao, Q.; Quan, X.; Chen, G. Novel phosphorus doped carbon nitride modified TiO_2 nanotube arrays with improved photoelectrochemical performance. *Nanoscale* **2015**, *7*, 16282–16289. [CrossRef] [PubMed]
34. Ran, J.; Ma, T.Y.; Gao, G.; Du, X.Y.; Qiao, S.Z. Porous P-doped graphitic carbon nitride nanosheets for synergistically enhanced visible-light photocatalytic H_2 production. *Energy Environ. Sci.* **2015**, *8*, 3708–3717. [CrossRef]
35. Datta, K.K.R.; Reddy, B.V.S.; Ariga, K.; Vinu, A. Gold nanoparticles embedded in a mesoporous carbon nitride stabilizer for highly efficient three-component coupling reaction. *Angew. Chem. Int. Ed.* **2010**, *49*, 5961–5965. [CrossRef] [PubMed]
36. Yuliati, L.; Yang, J.H.; Wang, X.; Maeda, K.; Takata, T.; Antonietti, M.; Domen, K. Highly active tantalum (v) nitride nanoparticles prepared from a mesoporous carbon nitride template for photocatalytic hydrogen evolution under visible light irradiation. *J. Mater. Chem.* **2010**, *20*, 4295–4298. [CrossRef]

Article

Reduction of Nitroarenes into Aryl Amines and *N*-Aryl hydroxylamines via Activation of $NaBH_4$ and Ammonia-Borane Complexes by Ag/TiO_2 Catalyst

Dimitrios Andreou [1], Domna Iordanidou [1], Ioannis Tamiolakis [2], Gerasimos S. Armatas [2] and Ioannis N. Lykakis [1,*]

1 Department of Chemistry, Aristotle University of Thessaloniki, University Campus, Thessaloniki 54124, Greece; dandreou@chem.auth.gr (D.A.); diordani@chem.auth.gr (D.I.)

2 Department of Materials Science and Technology, University of Crete, Vassilika Vouton, Heraklion 71003, Greece; gtam@materials.uoc.gr (I.T.); garmatas@materials.uoc.gr (G.S.A.)

* Correspondence: lykakis@chem.auth.gr; Tel./Fax: +30-2310-997-871

Academic Editors: Hermenegildo García and Sergio Navalón
Received: 10 February 2016; Accepted: 10 March 2016; Published: 22 March 2016

Abstract: In this study, we report the fabrication of mesoporous assemblies of silver and TiO_2 nanoparticles (Ag/MTA) and demonstrate their catalytic efficiency for the selective reduction of nitroarenes. The Ag/TiO_2 assemblies, which show large surface areas (119–128 $m^2 \cdot g^{-1}$) and narrow-sized mesopores (*ca.* 7.1–7.4 nm), perform as highly active catalysts for the reduction of nitroarenes, giving the corresponding aryl amines and *N*-aryl hydroxylamines with $NaBH_4$ and ammonia-borane (NH_3BH_3), respectively, in moderate to high yields, even in large scale reactions (up to 5 mmol). Kinetic studies indicate that nitroarenes substituted with electron-withdrawing groups reduced faster than those with electron-donating groups. The measured positive ρ values from the formal Hammett-type kinetic analysis of *X*-substituted nitroarenes are consistent with the proposed mechanism that include the formation of possible [Ag]-H hybrid species, which are responsible for the reduction process. Because of the high observed chemo selectivities and the clean reaction processes, the present catalytic systems, *i.e.*, Ag/MTA-$NaBH_4$ and Ag/MTA-NH_3BH_3, show promise for the efficient synthesis of aryl amines and *N*-aryl hydroxylamines at industrial levels.

Keywords: silver nanoparticles; nitroarenes; aryl amines; *N*-aryl hydroxylamines; titania; heterogeneous catalysis; selective reduction

---

1. Introduction

Noble metal nanoparticles are well known for their novel applications in the field of catalysis, biotechnology, bio-engineering and environmental remediation [1–5]. In recent years, the synthesis of noble metal nanoparticles as well as their applications, especially in catalysis, is of great interest for further study. Among them, silver nanoparticles (AgNPs) have attracted considerable attention because of their low cost, strong plasmonic properties and superior catalytic activity [6–8]. Several recent efforts to improve the physical properties of AgNPs have focused on synthesis of size- and shape-controlled nanoparticles, which were intended to enhance their catalytic and biological performance [6–11]. However, the practical use of individual AgNPs is often hampered by their severe aggregation during the catalytic reactions. In general, there are two main synthetic routes to overcome these limitations and to obtain small AgNPs with large exposed surface area. One is the surface modification of nanoparticles with organic molecules or surfactants in order to stabilize them against agglomeration and the other is the deposition of nanoparticles on a high-surface-area solid support such as metaloxides (SiO_2, TiO_2, ZrO_2, *etc.*), modified carbons, graphene, or other porous materials. The latter seems to be the most preferred method to sustain the stability and catalytic activity of nanoparticles [6–8]. Ag loaded TiO_2 (Ag/TiO_2) nanocomposites have recently emerged as promising catalysts for use in organic synthesis, because of the combination between TiO_2 electronic and optical properties, and Ag catalytic activities in chemical and biological areas. To this end, a variety of approaches have been used to prepare TiO_2 supported Ag catalysts, for instance, by photo deposition, chemical deposition and conventional impregnation method [6–12]. In these studies, however, most of the Ag/TiO_2 materials are composed of random aggregates of TiO_2 and Ag nanoparticles [6–8]. Moreover, the TiO_2 nanoparticles tend to agglomerate into bulk structures leading to a significant decrease in surface area and, therefore, in catalytic activity during repeated reactions.

Thus far, AgNPs have been successfully used as catalysts for several organic transformations, including C–C and C–N cross coupling reactions, cycloaddition and oxidative cyclization reactions, three-component reactions, oxidation of hydrosilanes to silanols, and reduction of imines to the corresponding amines [6–11]. In addition, there are several reports for the synthesis of AgNPs supported on different surfaces, which demonstrated their catalytic activity for the reduction of aromatic nitro compounds to the corresponding amines in the presence of sodium borohydride ($NaBH_4$) as reducing agent [13–22]. To our knowledge, only few examples of AgNPs-based catalyst exhibiting high catalytic activity in nitroarenes reduction have been reported, although excess of $NaBH_4$ or hast conditions (high temperature) are required [8,13–16]. In addition, most of these systems include reduction of nitrophenols by supported AgNPs in aqueous solution [8,17–22].

We recently reported the synthesis of mesoporous TiO_2 nanoparticle assemblies (MTA) using a one-pot chemical route. The MTA was prepared through sol-gel polymerization reaction between $TiCl_4$ and $Ti(OPr)_4$ in the presence of polyoxoethylene cetyl ether (Brij-58) block copolymer template [23,24]. Herein, we report the synthesis of new mesoporous hetero structures consisting of titania and silver nanoparticles (Ag/MTA) and demonstrate the catalytic activity of these materials towards the selective reduction of nitroarenes. The obtained Ag/TiO_2 assemblies show large internal surface area with narrow pore-size distribution and exhibit extraordinary activity for nitroarenes reduction. We present a systematic study of the reduction of several substituted nitroarenesto the corresponding anilines and *N*-aryl hydroxylamines, using $NaBH_4$ and ammonia-borane (NH_3BH_3) complexes as the reducing agents. Furthermore, a detailed mechanistic route for the AgNP-catalyzed reduction of nitroarenes in the presence of $NaBH_4$ or NH_3BH_3 is interpreted on the basis of kinetic analysis, as well as on the products identification by liquid chromatography-mass spectrometry (LC-MS) and nuclear magnetic resonance (NMR) spectroscopy.

2. Results and Discussion

2.1. Structural Properties of Ag/TiO_2 Catalysts

In this work, we employed commercial available $AgNO_3$ and AgOTf compounds, Degussa P (25) nanoparticles and mesoporous Ag-loaded TiO_2 nanoparticle assemblies (Ag/MTA) as catalysts for the selective reduction of various nitro compounds. All commercial catalysts were used as received. Mesoporous Ag/MTA composites with AgNPs loading amounts of 2, 3, 4 and 7 wt % were prepared by photochemical deposition of AgNPs on the surface of nanoparticle-based mesoporous titania (MTA) [24] (see Supplementary Materials for details). Energy dispersive X-ray spectroscopy (EDS) analysis of the obtained products evidenced that the Ag loadings in the Ag/MTA composites are very close to those expected from the stoichiometry of the reactions (see Table S1, Supplementary Materials).

The mesostructure and crystallinity of the Ag/MTA materials were characterized by X-ray diffraction (XRD), transmission electron microscopy (TEM) and nitrogen physisorption measurements. The powder XRD patterns (Figure S1, Supplementary Materials) indicated the well-defined crystal structure of Ag/MTA. They show a series of intense diffraction peaks in the $2\theta = 20°–80°$ range, which can be assigned to the anatase structure of TiO_2 (JCPDS #21-1272). In addition, the XRD patterns of Ag-loaded samples, especially those containing a high Ag loading (>3%), display weak diffractions at

~44.3° (002), ~64.5° (022) and ~77.4° (113), indicating the formation of metallic silver (JCPDS #04-0783). Figure 1a depicts a representative TEM image of the 4% Ag/MTA sample, which is the most active catalyst studied in this work. It reveals that this material consists of a porous network, which is composed of connected TiO_2 and Ag nanoparticles. On the basis of this technique, the average size of TiO_2 particles was estimated to be *ca.* 7–8 nm, while the diameter of Ag particles was found to be ~3–4 nm. Note that the average diameters of the TiO_2 and Ag nanoparticles were estimated by counting more than 100 particles in several TEM images (Figure S2, Supplementary Materials). To investigate further the crystal structure of mesoporous network, high-resolution TEM (HRTEM) imaging and selected-area electron diffraction (SAED) were performed. The HRTEM images indicate the single-crystal structure of the constituting nanoparticles, showing well-resolved and continuous lattice fringes across the particles. The lattice fringes in Figure 1b,c can be readily assigned to the anatase TiO_2 and face-centered cubic (fcc) Ag structure, respectively, in agreement with the XRD results. From the SAED pattern, it appeared that the crystalline structure of 4% Ag/MTA is a mixture of anatase TiO_2 and cubic Ag. In agreement with XRD and HRTEM results, all the Debye–Scherrer diffraction rings can be reasonably assigned to the anatase phase of TiO_2 (marker by red curves), while the additional diffraction spots could be indexed as (200) and (311) reflections of the cubic Ag (marker by blue dotted cycles) (Figure 1d).

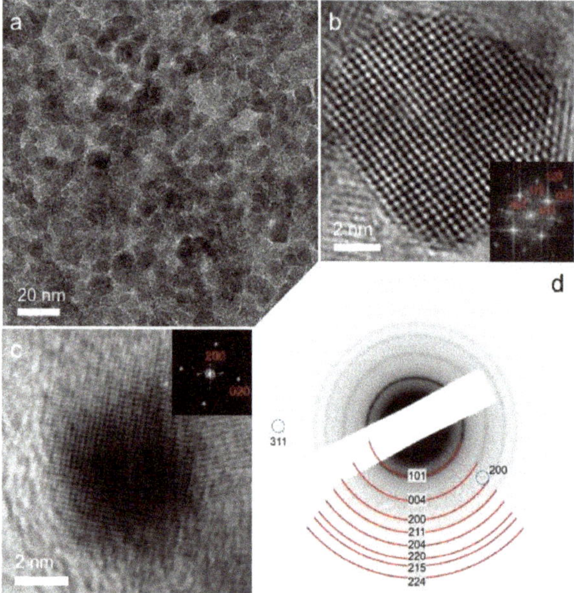

Figure 1. (**a**) Typical transmission electron microscopy (TEM) image; high resolution TEM (HRTEM) of a constituent (**b**) TiO_2 and (**c**) Ag nanoparticle; and (**d**) selected-area electron diffraction (SAED) pattern of mesoporous 4% Ag/mesoporous TiO_2 nanoparticle assemblies (MTA) catalyst. Insets of panels b and c: the corresponding fast Fourier transform (FFT) patterns indexed as the (111) and (100) zone axis diffraction of anatase TiO_2 and face-centered cubic Ag, respectively.

The mesoporosity of the Ag/MTA materials was probed with N_2 physisorption measurements at 77 K. The analysis showed that the Ag/MTA exhibit type-IV adsorption-desorption isotherms with H_2-type hysteresis loop (Figure S3, Supplementary Materials), which correspond to mesoporous materials with narrow-sized pores. The Brunauer-Emmett-Teller (BET) surface area and total pore volume of the Ag/MTA were measured to be 119–128 $m^2 \cdot g^{-1}$ and 0.21–0.23 $cm^3 \cdot g^{-1}$, respectively, which are slightly lower than those of the parent TiO_2 (MTA) sample (surface are ~149 $m^2 \cdot g^{-1}$ and pore volume ~0.27 $cm^3 \cdot g^{-1}$), probably due to the deposition of AgNPs on the TiO_2 surface. The pore diameter in Ag/MTA samples was obtained by fitting the adsorption isotherms using the non-local density functional theory (NLDFT) model (assuming slit-like pores), and was found to be ~7.1–7.4 nm with narrow pore-size distribution. Furthermore, a weak peak at around 5.4–5.6 nm was observed, especially for the high-Ag-loaded samples (>3 wt %), which could be attributed to the pores filling with AgNPs. Table S1 in the Supplementary Materials summarizes all the textural properties of mesoporous MTA and Ag/MTA materials.

2.2. *Evaluation of the Catalytic Activity*

Initially, we proceeded to optimize the catalytic conditions by studying the reduction of 4-nitrotoluene (1). In Table 1, we show the yields of 4-toluidine (**1a**) obtained using different Ag-containing catalysts, reducing agents and solvents. Specifically, the catalyst (10 mg) was placed in a 5 mL glass reactor, followed by the addition of solvent (1 mL), nitroarene (0.1 mmol) and hydride compound, and the reaction was vigorous stirred for appropriate time. To our delight, we observed that the 4% Ag/MTA catalyst with 6 mol-excess of $NaBH_4$ in ethanol affords quantitative conversion of **1** into 4-toluidine (**1a**) within 4 h (Table 1, entry 4).Remarkably, no by-products such as azoxy-, azo- or 1,2-diarylhydrazine were detected during the reaction progress by means of ^{1}H-NMR. In comparison, lower reduction yields of **1a** were obtained with Ag/MTA catalysts containing 7 wt % or less than 3 wt % Ag, as shown in Table 1, entries 2, 3 and 5 (see also Supplementary Materials, Figure S4). Meanwhile, TiO_2 alone, such as the MTA mesoporous and Degussa P25 nanoparticles (Table 1, entry 1) are largely inactive for the **1** reduction. Although, $AgNO_3$ and AgOTf catalyzed the conversion of **1** in high yields and short reaction time (Table 1, entries 6 and 7), an equimolar amount of the sesaltsis necessary for the reaction completion. In contrast, silver wire does not catalyze any reduction process (Table 1, entry 8). The reaction was incomplete with lower amounts of $NaBH_4$ (less than 2 mol-excess), while using 4 mol-excess of $NaBH_4$ a prolonged reaction time of 24 h was required to obtain **1a** in >91% yield (results not shown). In addition, no reduction of **1** was observed under mild conditions, for example, using 1 bar of H_2 at room temperature (Table 1, entry 9) or in the presence of transfer hydrogenation reagent such as the 1,1,3,3-tetramethyl disiloxane (TMDS) (Table 1, entry 10). In addition, with dimethylphenylsilane (DMPS), the conversion yield to the corresponding amine (**1a**) was only 11% (Table 1, entry 11). It is also noted that hydrazine hydrate ($NH_2NH_2 \cdot H_2O$) can be activated under our catalytic conditions proceeding to quantitative reduction of **1**, but higher temperature and prolonged reaction time are required (Table 1, entry 12). In contrast, when ammonia-borane (NH_3BH_3) is used as reducing agent (2.5 mol-excess based on **1**) the corresponding *N*-aryl hydroxylamine (**1b**) was detected by ^{1}H-NMR as the major product of **1** reduction, accompanying with small amount of amine **1a** (Table 1, entry 13). Control experiments showed no appreciable reduction of **1** in the absence of catalyst, indicating that the reduction process is catalytic in nature (result not shown). Finally, the reduction of **1** proceeded also efficiently in MeOH (Table 1, entry 14), while in other non-protic polar or apolar solvents no significant amount of **1a** was observed (Table 1, entries 15–19). Thus, 4% Ag/MTA catalyst in the presence of $NaBH_4$ inethanol affords fast, quantitative, and clean reduction of **1** without the requirement of any chromatographic purification of the product **1a** (Table 1, entry 5).After completion of **1**, as evidenced by thin layer chromatography (TLC), the slurry was filtered through a short pad of silica to withhold the catalyst with the aid of ethanol (~5 mL). Then, the filtrate was evaporated under reduced pressure to give pure 4-toluidine **1a** (98% isolated yield) as a brown solid. It is worth noting that our catalytic conditions are significantly milder(room temperature and 6-mol excess of $NaBH_4$) than those used in other studies, in which similar hydrogenation reactions were examined; however, either large excess of $NaBH_4$ (~25–800 mol excess), or hast conditions (>100 °C) are required [13–22]. Based on this, our catalytic system is highly feasible for practical use.

Table 1. Evaluation of various Ag-containing catalysts, reducing agents and solvents in the catalytic reduction of 4-nitrotoluene (**1**) into 4-toluidine (**1a**).

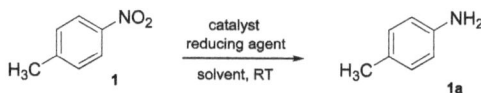

$$H_3C-C_6H_4-NO_2\ (\mathbf{1}) \xrightarrow[\text{solvent, RT}]{\text{catalyst, reducing agent}} H_3C-C_6H_4-NH_2\ (\mathbf{1a})$$

Entry	Catalyst [1]	Solvent	Reducing agent [2]	Time/Yield [3]
1	MTA or P25	EtOH	$NaBH_4$	24h/0%
2	2% Ag/MTA	EtOH	$NaBH_4$	4h/10% [4]
3	3% Ag/MTA	EtOH	$NaBH_4$	4h/51% [4]
4	**4% Ag/MTA**	**EtOH**	**$NaBH_4$**	**4h/>99%**
5	7% Ag/MTA	EtOH	$NaBH_4$	4h/30% [4]
6	AgOTf	EtOH	$NaBH_4$	0.5h/>99% [5]
7	$AgNO_3$	EtOH	$NaBH_4$	0.5h/>99% [5]
8	Ag (wire)	EtOH	$NaBH_4$	24h/0%
9	4% Ag/MTA	EtOH	H_2 (1 atm)	24h/0%
10	4% Ag/MTA	EtOH	TMDS	24h/0%
11	4% Ag/MTA	EtOH	DMPS	4h/11%
12	4% Ag/MTA	EtOH	$NH_2NH_2 \cdot H_2O$	24%/99% [6]
13	**4% Ag/MTA**	**EtOH**	**NH_3BH_3**	**0.2h/6% (94%)** [7]
14	4% Ag/MTA	MeOH	$NaBH_4$	4h/99%
15	4% Ag/MTA	Toluene	$NaBH_4$	4h/0%
16	4% Ag/MTA	THF	$NaBH_4$	4h/0%
17	4% Ag/MTA	CH_3CN	$NaBH_4$	4h/0%
18	4% Ag/MTA	DCM	$NaBH_4$	4h/0%
19	4% Ag/MTA	Acetone	$NaBH_4$	4h/0%

[1] Ten milligrams of each catalyst was used. [2] sixe mol-excess of $NaBH_4$ were used, while 1,1,3,3-tetramethyl disiloxane (TMDS), dimethylphenylsilane (DMPS) and hydrazine were added in 2.5, 5 and 12 fold-excess of mmols, respectively, based on **1**. [3] Relative yield of **1a** determined by ^{1}H-NMR. [4] The conversions of **1** were in the range of 75%–100%, while the azoxy-, azo- and hydrazo-arenes were formed as major products. [5] Using lower amount of the $AgNO_3$ and AgOTf (20% mol and 50% mol, based on **1**) the conversions of **1** were in the range of 10%–79% after 2h; however, the corresponding hydrazo- and azo-arenes were formed as the major products accompynaning with small amount of the amine **1**. [6] Twelve mol-excess of hydrazine were used for reaction completion at 90 °C. [7] The corresponding *N*-aryl hydroxylamine **1b** was formed as the major product accompanying with small amount of the amine **1a**, within 10 min.

2.3. *Chemoselective Reduction of Nitroarenes into Aryl Amines and N-Aryl Hydroxylamines*

To study the limitation of this chemoselective reduction process, a series of nitroarenes (**1–12**) were examined under the above described conditions. As shown in Scheme 1, 4% Ag/MTA catalyst in the presence of $NaBH_4$ (6 mol-excess based on nitroarene amount) produce the corresponding substituted anilines **1a–12a** in excellent yields (>90%) and selectivity (>98%). Moreover, bromo and chloro-substituted nitroarenes (**3** and **4**) were also reduced without undergoing any dehalogenation, while *p*-dinitrobenzene (**9**) was completely converted too diamine (**9a**) within 4 h. Similarly, carboxylate, and cyano functionalities in **5** and **8** nitroarenes remained intact under the examined conditions, indicating highly chemoselective reduction. Consistent with the above results, the reduction of 6-nitro-isobenzofuran-1(3*H*)-one (**10**) gave also the corresponding amine without further transformation of the lacton ring. Notably, 3-nitrostyrene **11** was reduced to the corresponding saturated amine **11a** in 70% relative yield, accompanying with a 3-ethyl-nitrobenzene yield of 30% (see Supplementary Materials). This result suggests that the present catalytic system is also active for the preferred hydrogenation of the C=C double bond relative to the nitro group of **11**.

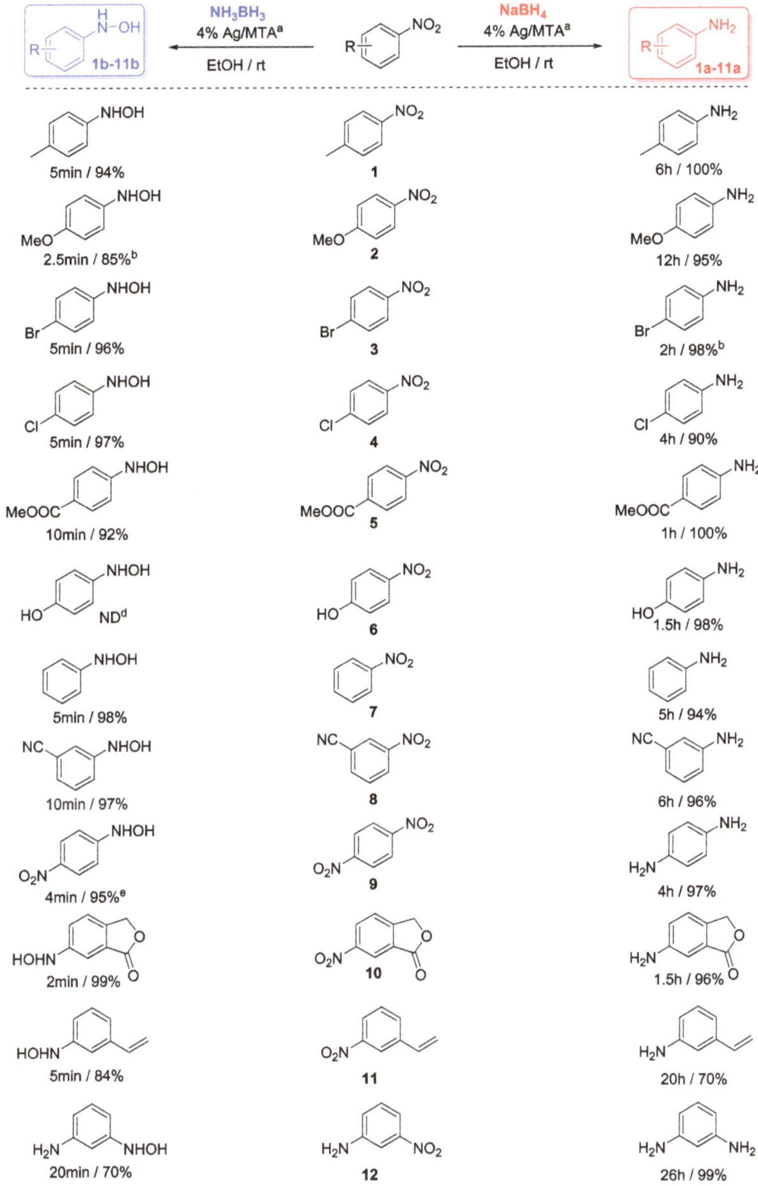

[a]Ten milligrams of each catalyst was used.In all cases quantitatevaly conversion of the corresponding nitroarene was observed and the percentages below each product correspond to its relative yield determined by ^{1}H-NMR. [b]Totally, 1.5 mol-excess of NH_3BH_3 was used based on **2** amount. [c]Twenty milligrams of the catalyst were used. [d]Not detected, even at initial reaction time (see also Supplementary Materials). [e]Totally, 2mol-excess of NH_3BH_3wasused, based onnitroarene **9** amount.

Scheme 1. Chemoselective reduction of nitroarenes (**1–12**) into aryl amines (**1a–12a**) and *N*-aryl hydroxylamines (**1b–12b**) catalyzed by 4% Ag/MTA with $NaBH_4$ and NH_3BH_3 complexes, respectively.

Nevertheless, when the NH_3BH_3 (1.5–2.5 mol-excess based on nitroarene amount) is used as the reducing agent, the corresponding *N*-aryl hydroxylamines (**1b–12b**) are formed under the above described conditions (Scheme 1); indeed, in high relative yields (>84%) and within short reaction times (2–10 min). ^{1}H-NMR analysis of the crude mixtures showed only a small amount of the corresponding substituted anilines (**1a–10a**) (2%–10%)as byproducts (see Supplementary Materials). As shown in Scheme 1, the reduction of *p*-nitrophenol (**6**) produces the corresponding amine **6a** as the only product. In this case, *N*-aryl hydroxylamine **6b** was not detected by NMR at initial reaction time (see Supplementary Materials, Figure S5). Reduction of **11** leads to the *N*-aryl hydroxylamine **11b** accompynaning with small amount of the initial nitroarene, however, under the described conditions **12** gave the **12b** in 70% relative yield with a significant amount of the corresponding amine **12a** (30%), (see Supplementary Materials). It should be noted that the reaction products were kept at room temperature and no further chromatographic purification was performed to avoid decomposition or transformation of the *N*-aryl hydroxylamines into the corresponding nitrosoarenes, anilines and azoxy- or azo-arenes. In the literature, several synthetic routes towards *N*-aryl hydroxylamines formation have been reported; however, certain limitations related to the applicability of these reactions (including scale-up synthesis) have been imposed. Since the first procedure using Zn/NH_4Cl [25] and the enzymatic nitroreductase system [26], the common synthetic routes associated with the title transformation have focused on the direct hydrogenation (with H_2) of nitroarenes using precious metal nanoparticles such as Pd, Pt, Ru, Re and Rh as catalysts [27–33]. In addition, catalytic transfer hydrogenation reactions of nitroarenes to *N*-aryl hydroxylamines were also realized using Pt/hydrazine [34], Zn in CO/H_2O [35] and Sb/$NaBH_4$ [36] systems, while only recently Au/TiO_2 nanoparticles have been employed for the selective nitroalkanes reduction to *N*-alkyl hydroxylamines [37]. To our knowledge, since now, heterogeneous Ag-catalyzed hydrogenation of nitroarenes to the corresponding *N*-aryl hydroxylamines is an unknown transformation. These results, accompanying with the observation that *N*-aryl hydroxylamines are formed in high relative yields and through afast and pure reaction process, suggest that the present catalytic system Ag/MTA-NH_3BH_3 can be applicable to various hydrogenation reactions, including fine synthesis of *N*-aryl hydroxylamines.

2.4. Kinetic Studies

The reusability of the 4% Ag/MTA was examined by conducting repeat catalytic experiments. The 4% Ag/MTA catalyst can be easily separated from the reaction mixture by simply filtration and it can be reused for the next catalytic run. As shown in Supplementary Materials, Figure S6, the catalyst can be used at least three times without significant loss of its catalytic activity and selectivity. In order to assess the feasibility of the present catalytic system, the 4% Ag/MTA catalyst was further tested for large-scale production of aryl amines and *N*-aryl hydroxylamines from nitroarenes. For this reason, 5 mmol of nitroarene **1** were reduced in the presence of 4% Ag/MTA (0.8 mol %) with 8 mol-excess of $NaBH_4$ in 15 mL EtOH. After completion of the reaction (within ~20 h based on TLC analysis), the mixture was filtered upon silica gel, washed with ethanol and purified by column chromatography to afford the corresponding 4-toluedine **1a** in 91% isolated yield. As a comparison, under the same reaction conditions but with 3 mol-excess of NH_3BH_3, the corresponding *N*-aryl hydroxyamine **1b** was obtained at 93% relative yield in 30 min, according to the ^{1}H-NMR analysis of the crude product (results not shown). These results correspond to a turn over number (TON) of about 125 and a turnover frequency (TOF) of 250 h^{-1}.

To propose a plausible mechanistic pathway for the present Ag-catalyzed reduction of nitroarenes, we performed a Hammett-type kinetic study for the reduction of a diverse set of *para*- and meta-X-substituted-nitroarenes (**1**, **2**, **3**, **4**, **5**, **7**, **8** and **12**). The kinetic studies were carried out as follow: 0.2 mmol of the nitroarene, 20 mg of 4% Ag/MTA and 0.6 mmol of $NaBH_4$ were added in ethanol (1 mL). A 100 μL aliquot of the mixture was taken at appropriate time and the mixture was filtrated through a short pad of silica (to withhold the catalyst) and washed with ethanol (~1 mL). Then, the filtrate was evaporated under reduced pressure and the consumptions of the corresponding X-substituted nitroarene were determined by integrating the appropriate proton signals in ^{1}H-NMR spectra. Each reaction was repeated at least three times and the average values are depicted in Figure 2 and in the Supplementary Materials, Figures S7 and S8. Considering that the concentration of the possible formed silver-hydride species remains constant at initial times and assuming a pseudo-first order dependence of the reaction rate on the nitroarene concentration, Equation (1) can be applied.

$$\ln(x) = -kt \quad (1)$$

where, k is the rate constant and x is the consumption of the X-substituted nitroarene at reaction time t.

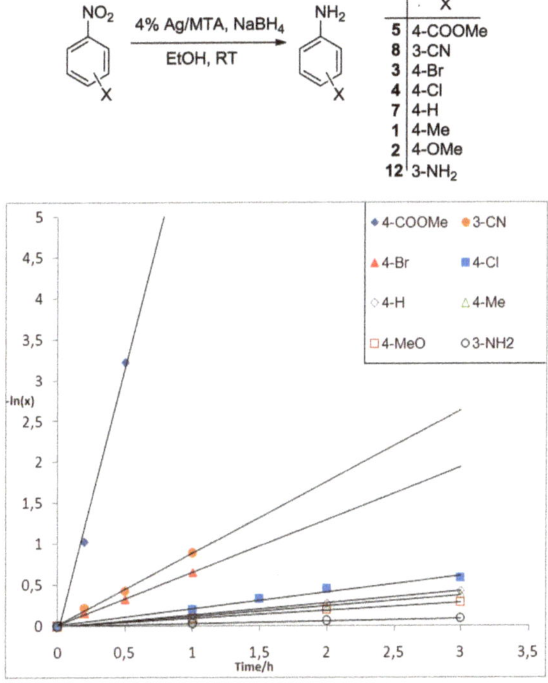

Figure 2. Kinetic analysis of the 4% Ag/MTA-catalyzed reduction of various X-substituted nitroarenes (X = 3-NH_2 (**12**), 4-MeO (**2**), 4-Me (**1**), 4-Br (**3**), 4-H (**7**), 4-Cl (**4**), 3-CN (**8**) and 4-COOMe (**5**)) with $NaBH_4$.

According to Equation (1), a plot of the $\ln(x)$ *versus* time gives a linear curve, the slope of which is equal to the rate constant k. The results indicated that the kinetic activity of **1**, **2**, **3**, **4**, **5**, **7**, **8** and **12** nitroarenes is remarkably affected by the nature of the X-substituent group, in which the reduction proceeds faster as the electron-withdrawing ability of the substituent group increases. For example, the reduction of **3** (4-Br), **4** (4-Cl), **5** (4-COOMe) and **8** (3-CN) proceeds in a faster rate than that with the nitrobenzene **7** (X = H) (Figure 2), as indicated by the relative rate constant ratios $k_{COOMe}/k_H = 25$, $k_{CN}/k_H = 6.0$, $k_{Br}/k_H = 4.5$ and $k_{Cl}/k_H = 1.4$, respectively (see also Supplementary Materials, Figures S7 and S8). However, nitroarenes containing electron-donating group, such as **12** (3-NH_2), **2** (4-MeO) and **1** (4-Me), were reduced with slower reaction rate (Figure 2); the relative rate constant rations were $k_{NH2}/k_H = 0.2$, $k_{MeO}/k_H = 0.7$ and $k_{Me}/k_H = 0.9$, respectively. In addition to these results, a Hammett-type correlation in the competition of X-substituted nitroarenes (**12**, **2**, **1**, **4**, **3**, **8** and **5**) *versus* nitrobenzene (**7**) gave positive slopes, *i.e.*, $\rho \approx 0.8$, $R^2 = 0.890$ (using σ^+ values) and $\rho \approx 0.9$, $R^2 = 0.826$ (using σ values) (Supplementary Materials, Figure S9). The small ρ values for these correlations indicate that the reaction mechanism involves either radical intermediates or a transition state with a small charge separation [38]. However, the measured positive ρ values, are consistent with a proposed mechanism that include a negative charge (or hydride transfer) in the transition state, which is stabilized by electron-withdrawing substituents [38,39].

According to the above results, we propose a general mechanistic pathway for the AgNPs catalyzed reduction of nitroarenes. First, a B-H bond cleavage occurs to gives [Ag]-H active species. Such hybrid species are responsible for the rapid reduction of nitroarenes into the corresponding *N*-aryl hydroxylamines. This assumption is consistent with the kinetics studies showing that nitroarenes bearing electron-withdrawing substituents reduced faster than those with electron donating groups. Finally, *N*-aryl hydroxylamines are further reduced into the corresponding aryl amines with $NaBH_4$; however, this step becomes slower in the presence of NH_3BH_3. To shed light on the above hypothesis, the catalytic reductions of electron-donating **2** and electron-withdrawing **5** and **9** *p*-*X*-substituted nitroarenes were also separately conducted in CD_3OD using 4% Ag/MTA. Each reduction process was monitored directly by ^{1}H-NMR spectroscopy, at initial reaction times. As shown in Figures S10–S12 of the Supplementary Materials, during the reduction process, the only intermediate products were the corresponding *N*-aryl hydroxylamines **2b**, **5b** and **9b**.

3. Experimental Section

3.1. Materials

Brij 58 surfactant ($HO(CH_2CH_2O)_{20}C_{16}H_{33}$, Mn~1124), titanium tetrachloride (99.9%), $AgNO_3$ (>99%), AgOTf, absolute ethanol (99.8%), $NaBH_4$, NH_3BH_3, TMDS and DMPS were purchased from Sigma-Aldrich (Darmstadt, Germany). Titanium(IV) isopropoxide (>98%) was purchased from Merck (Darmstadt, Germany). TiO_2 nanoparticles (P25) were purchased from Degussa AG (Dusseldorf, Germany).The aromatic nitro compounds used as substrates were of high purity and commercially available from Aldrich (Darmstadt, Germany).

3.2. Synthesis of Ag/MTA Catalysts

The mesoporous TiO_2 nanoparticle assemblies (MTA) were prepared according to the method reported previous [24]. Ag-loaded TiO_2 (Ag/MTA) catalysts with different loading of AgNPs were obtained by photocatalytic reduction method. Typically, 0.2 g of MTA were dispersed into 10 mL of a CH_3CN/H_2O/ethanol (10:1:1 v/v) solution containing appropriate amounts of $AgNO_3$. The suspension was then illuminated with a 5 mW ultraviolet lamp (λ = 365 nm) for 2 h under continuous stirring. The product was then collected by filtration, washed with ethanol, and dried at 60 °C for 12 h. A series of mesoporous Ag/MTA catalysts with different Ag loadings, *i.e.*, x = 2, 3, 4 and 7 wt %, was prepared using 6.4, 9.6, 13.2 and 23.8 mg of $AgNO_3$, respectively.

3.3. Physical Characterization

The X-ray diffraction (XRD) patterns were collected using a Panalytical X'Pert Pro MPD X-ray diffractometer (45 kV and 40 mA, Lelyweg, the Netherlands) with a Cu Kα radiation (λ = 1.5406 Å). Elemental microprobe analysis was performed on a JEOL Model JSM-6390LV scanning electron microscopy (SEM, Tokyo, Japan) system equipped with an Oxford INCA PentaFET-x3 energy-dispersive X-ray spectroscopy (EDS) detector (Oxfordshire, UK). Data acquisition was performed several times using an accelerating voltage of 20 kV and 60 s accumulation time. Transmission electron microscopy (TEM) experiments were carried out with a JEOL model JEM-2100 electron microscope (LaB_6 filament) operating at 200 kV. The sample was dispersed in ethanol by sonication, and the dispersion was then dropped onto a Cu grid covered with carbon film. Nitrogen adsorption-desorption isotherms were measured at liquid N_2 temperature (77 K) on a NOVA 3200e volumetric analyzer (Quantachrome, Boynton Beach, FL, USA). Before analysis, samples were degassed overnight at 150 °C under vacuum (<10^{-5} Torr) to remove moisture. The specific surface areas were calculated using the Brumauer-Emmett-Teller (BET) method [40] on the adsorption data in the 0.06–0.25 relative pressure (P/P_o) range. The total pore volumes were derived from the adsorbed volume at P/P_o = 0.99 and the pore size distributions were obtained by the nonlocal density functional theory (NLDFT) method [41] based on the adsorption data.

3.4. Catalytic Reactions

Supported silver catalyst Ag/MTA (10 mg) was placed in a 5 mL glass reactor, followed by the addition of ethanol (1 mL), nitro compound (0.1 mmol) and $NaBH_4$ (0.6 mmol) or NH_3BH_3 (0.25 mmol), and the reaction mixture was stirred at room temperature for a selected time. The reaction was monitored by thin layer chromatography (TLC), and after completion, the slurry was filtered under pressure through a short pad of silica to withhold the catalyst with the aid of ethanol or methanol (~5 mL). After solvent evaporation the corresponding products were formed in pure forms. Product analysis was conducted by ^{1}H-NMR and ^{13}C-NMR spectroscopy (Bruker AM 300, Bruker Biospin GMBH, Rheinstetten, Germany and Agilent AM 500, Agilent Technologies, Santa Clara, CA, USA). Identification of the products was realized by comparing the NMR spectra with those of the commercially available pure substances. LC-MS 2010 EV Instrument (Shimadzu, Tokyo, Japan) under Electrospray Ionization (ESI) conditions was used for the determination of the mass spectra.

Reusability testing of the catalyst was conducted in the case of the 4% Ag/MTA sample through the reduction of 4-nitrotoluene (1). A 2 mL mixture of the feeding solution (0.2 mmol of **1**, 1.2 mmol of $NaBH_4$ and 20 mg of catalyst (4 mol % Ag)) was placed into a vial. Each catalytic reaction was stopped after 6 h and the catalyst was collected by filtration, washed with ethanol and dried in an oven at 100 °C for 12 h. Then, the recovered catalyst was used for the next catalytic run without any additional treatment.

4. Conclusions

In conclusion, we have shown that mesoporous titania supported silver nanoparticles (Ag/MTA) can effectively catalyze the chemoselective reduction of nitroarenes into the corresponding aryl amines and *N*-aryl hydroxylamines, employing $NaBH_4$ and NH_3BH_3 as reducing agents, respectively. Product analysis and kinetic studies indicated that aryl amine formation proceeds through a reduction pathway involving the initial formation of silver-hydride species; although additional mechanistic studies are required in this direction. In both catalytic processes, the corresponding *N*-aryl hydroxylamines were observed either as intermediates (with $NaBH_4$) or as the major products (with NH_3BH_3). Based on the observed high chemoselectivities and the fast and clean reaction processes, both catalytic systems, Ag/MTA-$NaBH_4$ and Ag/MTA-NH_3BH_3, can be applicable to various hydrogenation reactions, including fine synthesis of amines and *N*-aryl hydroxylamines, respectively.

Supplementary Materials: They are available online at http://www.mdpi.com/2079-4991/6/3/54/s1.

Acknowledgments: Financial supports by the European Union and the Greek Ministry of Education (ERC-09 and ARISTEIA-2691) are kindly acknowledged. I.N.L. gratefully acknowledges the sponsorship from COST action CM1201. We thank E. Evgenidou for obtaining the MS spectra.

Author Contributions: D.A., D.I. and I.T. designed and performed all the experiments and interpretation of results. G.S.A. and I.N.L supervised this work, prepared and reviewed the manuscript.

Conflicts of Interest: The authors declare no conflict of interest.

References

1. Chaudhuri, R.G.; Paria, S. Core/shell nanoparticles: Classes, properties, synthesis mechanisms, characterization, and applications. *Chem. Rev.* **2012**, *112*, 2373–2433. [CrossRef] [PubMed]
2. Sreeprasad, T.S.; Pradeep, T. Noble metal nanoparticles. In *Springer Handbook of Nanomaterials*; Vajtai, R., Ed.; Springer-Verlag: Berlin, Germany, 2013; pp. 303–388.
3. Astruc, D. Transition-metal nanoparticles in catalysis: From historical background to the state-of-the art. In *Nanoparticles and Catalysis*; Astruc, D., Ed.; Wiley-VCH Verlag GmbH and Company KGaA: Weinheim, Germany, 2008; pp. 1–48.

4. Doria, C.; Conde, J.; Veigas, B.; Giestas, L.; Almeida, C.; Assunção, M.; Rosa, J.; Baptista, P.V. Noble metal nanoparticles for biosensing applications. *Sensor* **2012**, *12*, 1657–1687. [CrossRef] [PubMed]
5. De, M.; Ghosh, P.S.; Rotello, V.M. Applications of nanoparticles in biology. *Adv. Mater.* **2008**, *20*, 4225–4241. [CrossRef]
6. El-Nour, K.M.M.A.; Eftaiha, A.; Al-Warthan, A.; Ammar, R.A.A. Synthesis and applications of silver nanoparticles. *Arabian J. Chem.* **2010**, *3*, 135–140. [CrossRef]
7. Tran, Q.H.; Nguyen, V.Q.; Le, A.-T. Silver nanoparticles: Synthesis, properties, toxicology, applications and perspectives. *Adv. Nat. Sci. Nanosci. Nanotechnol.* **2013**, *4*. [CrossRef]
8. Bhosale, M.A.; Bhanage, B.M. Silver nanoparticles: Synthesis, characterization and their application as a sustainable catalyst for organic transformations. *Curr. Org. Chem.* **2015**, *19*, 708–727. [CrossRef]
9. Abbiati, G.; Rossi, E. Silver and gold-catalyzed multicomponent reactions. *Beilstein J. Org. Chem.* **2014**, *10*, 481–513. [CrossRef] [PubMed]
10. Rycnge, M.; Cobley, C.M.; Zeng, J.; Li, W.; Moran, C.H.; Zhang, Q.; Qin, D.; Xia, Y. Controlling the synthesis and assembly of silver nanostructures for plasmonic applications. *Chem. Rev.* **2011**, *111*, 3669–3712.
11. Dong, X.-Y.; Gao, Z.-W.; Yang, K.-F.; Zhang, W.-Q.; Xu, L.-W. Nanosilver as a new generation of silver catalysts in organic transformations for efficient synthesis of fine chemicals. *Catal. Sci. Technol.* **2015**, *5*, 2554–2574. [CrossRef]
12. Kundu, S.; Mandal, M.; Ghosh, S.K.; Pal, T. Photochemical deposition of SERS active silver nanoparticles on silica gel and their application as catalysts for the reduction of aromatic nitro compounds. *J. Colloid Interface Sci.* **2004**, *272*, 134–144. [CrossRef] [PubMed]
13. Davarpanah, J.; Kiasat, A.R. Catalytic application of silver nanoparticles immobilized to rice husk-SiO_2-aminopropylsilane composite as recyclable catalyst in the aqueous reduction of nitroarenes. *Catal. Commun.* **2013**, *41*, 6–11. [CrossRef]
14. Solanki, J.N.; Murthy, Z.V.P. Reduction of nitro aromatic compounds over Ag/Al_2O_3 nanocatalyst prepared in water-in-oil microemulsion: Effects of water-to-surfactant mole ratio and type of reducing agent. *Ind. Eng. Chem. Res.* **2011**, *50*, 7338–7344. [CrossRef]
15. Zhou, Q.; Qian, G.; Li, Y.; Zhao, G.; Chao, Y.; Zheng, J. Two-dimensional assembly of silver nanoparticles for catalytic reduction of 4-nitroaniline. *Thin Solid Films* **2008**, *516*, 953–956. [CrossRef]
16. Pradhan, N.; Pal, A.; Pal, T. Silver nanoparticle catalyzed reduction of aromatic nitro compounds. *Colloids Surf. A* **2002**, *196*, 247–257. [CrossRef]
17. Dong, Z.; Le, X.; Li, X.; Zhang, W.; Dong, C.; Ma, J. Silver nanoparticles immobilized on fibrous nano-silica as highly efficient and recyclable heterogeneous catalyst for reduction of 4-nitrophenol and 2-nitroaniline. *Appl. Catal. B* **2014**, *158*, 129–135. [CrossRef]
18. Chi, Y.; Tu, J.; Wang, M.; Li, X.; Zhao, Z. One-pot synthesis of ordered mesoporous silver nanoparticle/carbon composites for catalytic reduction of 4-nitrophenol. *J. Colloid Interface Sci.* **2014**, *423*, 54–59. [CrossRef] [PubMed]
19. Deshmukh, S.P.; Dhokale, R.K.; Yadav, H.M.; Achary, S.N.; Delekar, S.D. Titania-supported silver nanoparticles: An efficient and reusable catalyst for reduction of 4-nitrophenol. *Appl. Surf. Sci.* **2013**, *273*, 676–683. [CrossRef]
20. Wang, M.; Tian, D.; Tian, P.; Yuan, L. Synthesis of micron-SiO_2@nano-Ag particles and their catalyticperformance in 4-nitrophenol reduction. *Appl. Surf. Sci.* **2013**, *283*, 389–395. [CrossRef]
21. Mohamed, M.M.; Al-Sharif, M.S. One pot synthesis of silver nanoparticles supported on TiO_2 using hybrid polymers as template and its efficient catalysis for the reduction of 4-nitrophenol. *Mater. Chem. Phys.* **2012**, *136*, 528–537. [CrossRef]
22. Jiang, H.-L.; Akita, T.; Ishida, T.; Haruta, M.; Xu, Q. Synergistic catalysis of Au@Ag core-shell nanoparticles stabilized on metal-organic framework. *J. Am. Chem. Soc.* **2011**, *133*, 1304–1306. [CrossRef] [PubMed]
23. Tamiolakis, I.; Fountoulaki, S.; Vordos, N.; Lykakis, I.N.; Armatas, G.S. Mesoporous Au-TiO_2 nanoparticle assemblies as efficient catalysts for the chemoselective reduction of nitro compounds. *J. Mater. Chem. A* **2013**, *1*, 14311–14319. [CrossRef]
24. Tamiolakis, I.; Lykakis, I.N.; Katsoulidis, A.P.; Armatas, G.S. One-pot synthesis of highly crystalline mesoporous TiO_2 nanoparticle assemblies with enhanced photocatalytic activity. *Chem. Commun.* **2012**, *48*, 6687–6689. [CrossRef] [PubMed]
25. Kamm, O. β-phenylhydroxylamines. *Org. Synth.* **1925**, *4*, 57–58.

26. Nguyen-Tran, H.-H.; Zheng, G.-W.; Qian, X.-H.; Xu, J.-H. Highly selective and controllable synthesis of arylhydroxylamines by the reduction of nitroarenes with an electron-withdrawing group using a new nitroreductase *Ba*NTR1. *Chem. Commun.* **2014**, *50*, 2861–2864. [CrossRef] [PubMed]
27. Boymans, E.H.; Witte, P.T.; Vogt, D. A study on the selective hydrogenation of nitroaromatics to *N*-arylhydroxylamines using a supported Pt nanoparticle catalyst. *Catal. Sci. Technol.* **2015**, *5*, 176–183. [CrossRef]
28. Rong, Z.; Du, W.; Wang, Y.; Lu, L. Carbon supported Pt colloid as effective catalyst for selective hydrogenation of nitroarenes to arylhydroxylamines. *Chem. Commun.* **2010**, *46*, 1559–1561. [CrossRef] [PubMed]
29. Takenaka, Y.; Kiyosu, T.; Choi, J.C.; Sakakura, T.; Yasuda, H. Selective synthesis of *N*-aryl hydroxylamines by the hydrogenation of nitroaromatics using supported platinum catalysts. *Green Chem.* **2009**, *11*, 1385–1390. [CrossRef]
30. Pernoud, L.; Candy, P.; Didillon, B.; Jacquot, R.; Basset, J.M. *Studies in Surface Science and Catalysis*; Avelino Corma, F.V.M.S.M., José Luis, G.F., Eds.; Elsevier: Amsterdam, the Netherlands, 2000; Volume 130, pp. 2057–2062.
31. Tamura, M.; Kon, K.; Satsuma, A.; Shimizu, K. Volcano-curves for dehydrogenation of 2-propanol and hydrogenation of nitrobenzene by SiO_2-supported metal nanoparticles catalysts as described in terms of a d-band model. *ACS Catal.* **2012**, *2*, 1904–1909. [CrossRef]
32. Widegren, J.A.; Finke, R.G. A review of soluble transition-metal nanoclusters as arene hydrogenation catalysts. *J. Mol. Catal. A* **2003**, *191*, 187–207. [CrossRef]
33. Karwa, S.L.; Rajadhyaksha, R.A. Selective catalytic hydrogenation of nitrobenzene to phenylhydroxylamine. *Ind. Eng. Chem. Res.* **1987**, *26*, 1746–1750. [CrossRef]
34. Shila, A.K.; Das, P. Solid supported platinum(0) nanoparticles catalyzed chemo-selective reduction of nitroarenes to *N*-arylhydroxylamines. *Green Chem.* **2013**, *15*, 3421–3428. [CrossRef]
35. Liu, S.; Wang, Y.; Jianga, J.; Jina, Z. The selective reduction of nitroarenes to *N*-arylhydroxylamines using Zn in a CO_2/H_2O system. *Green Chem.* **2009**, *11*, 1397–1400. [CrossRef]
36. Ren, P.; Dong, T.; Wu, S. Synthesis of *N*-arylhydroxylamines by antimony-catalyzed reduction of nitroarenes. *Synth. Commun.* **1997**, *27*, 1547–1552. [CrossRef]
37. Vasilikogiannaki, E.; Gryparis, C.; Kotzabasaki, V.; Lykakis, I.N.; Stratakis, M. Facile reduction of nitroarenes into anilines and nitroalkanes into hydroxylamines via the rapid activation of ammonia-borane complex by supported gold nanoparticles. *Adv. Synth. Catal.* **2013**, *355*, 907–911. [CrossRef]
38. Lowry, T.H.; Richardson, K.S. *Mechanism and Theory in Organic Chemistry*, 3rd ed.; Harper & Row: New York, NY, USA, 1987; pp. 60–71.
39. Fountoulaki, S.; Daikopoulou, V.; Gkizis, P.L.; Tamiolakis, I.; Armatas, G.S.; Lykakis, I.N. Mechanistic studies of the reduction of nitroarenes by $NaBH_4$ or hydrosilanes catalyzed by supported gold nanoparticles. *ACS Catal.* **2014**, *4*, 3504–3511. [CrossRef]
40. Brunauer, S.; Deming, L.S.; Deming, W.E.; Teller, E. On a theory of the van der Waals adsorption of gases. *J. Am. Chem. Soc.* **1940**, *62*, 1723–1732. [CrossRef]
41. Ravikovitch, P.I.; Wei, D.; Chueh, W.T.; Haller, G.L.; Neimark, A.V. Evaluation of pore structure parameters of MCM-41 catalyst supports and catalysts by means of nitrogen and argon adsorption. *J. Phys. Chem. B* **1997**, *101*, 3671–3679. [CrossRef]

Article

Synthesis of Ball-Like Ag Nanorod Aggregates for Surface-Enhanced Raman Scattering and Catalytic Reduction

Wenjing Zhang [1], Yin Cai [1], Rui Qian [1], Bo Zhao [2] and Peizhi Zhu [1,*]

[1] School of Chemistry and Chemical Engineering, Yangzhou University, Jiangsu 225002, China; 15262236582@163.com (W.J.Z.); yincai1992@sina.com (Y.C.); ruiqian2016@sina.com (R.Q.)

[2] Jiangsu Collaborative Innovation Center of Biomedical Functional Materials and Jiangsu Key Laboratory of Biofunctional Materials, School of Chemistry and Materials Science, Nanjing Normal University, Nanjing 210023, China; zhaobo@njnu.edu.cn

* Correspondence: pzzhu@yzu.edu.cn; Tel./Fax: +86-514-87975244

Academic Editors: Hermenegildo García and Sergio Navalón
Received: 30 March 2016; Accepted: 18 May 2016; Published: 27 May 2016

Abstract: In this work, ball-like Ag nanorod aggregates have been synthesized via a simple seed-mediated method. These Ag mesostructures were characterized by scanning electron microscope (SEM), transmission electron microscopy (TEM), ultraviolet-visible spectroscopy (UV-Vis), and X-ray diffraction (XRD). Adding a certain amount of polyvinyl pyrrolidone (PVP) can prolong its coagulation time. These Ag nanorod aggregates exhibit effective SERS effect, evaluated by Rhodamine 6G (R6G) and doxorubicin (DOX) as probe molecules. The limit of detection (LOD) for R6G and DOX are as low as 5×10^{-9} M and 5×10^{-6} M, respectively. Moreover, these Ag nanorod aggregates were found to be potential catalysts for the reduction of 4-nitrophenol (4-NP) in the presence of $NaBH_4$.

Keywords: Ag nanorod aggregates; surface-enhanced Raman scattering; Rhodamine 6G; doxorubicin; PVP; catalytic reduction; 4-nitrophenol

---

1. Introduction

In recent years, silver nanoparticles (AgNPs) have seen broad application in areas such as catalysis [1], biomedicine [2], antimicrobial agent [3] and SERS [4,5]. As a powerful molecular fingerprinting technique, surface-enhanced Raman scattering (SERS) is a sensitive technique for trace detection [6–10]. Noble metal nanoparticles such as Ag and Au particles have been extensively explored due to their high SERS-active properties [11–17]. It is well-established that SERS activities are size and shape dependent [18,19]. Ag nanoparticles with complex topography have more hot-spots on surface to amplify Raman scattering of probe molecules [20,21].

Silver nanoparticles have also gained much attention for their application as a sustainable catalyst for organic transformations owing to their unique electronic properties and high surface area to volume ratio [22]. Particularly, silver nanoparticles show highly efficient catalytic activity in oxidation of methanol and ethylene [23,24], as well as reduction of nitric oxides (NO_x) [25]. Yang *et al.* [26] proposed that the flower-like Ag microcrystal exhibited high catalytic activity for 4-nitrophenol reduction due to their high surface area and the local electromagnetic field intensity enhancement.

Many researchers have studied numerous Ag complex structures as highly sensitive SERS substrates and catalysts [27–31]. Various methods such as chemical reduction [32,33], template process [34], and galvanic replacement [35] have been used to synthesize functional Ag nanoparticles. Using a double-reductant approach, seed-mediated method has been explored to prepare Ag nanocubes [36,37], nanowires [38], nanopolyhedron [39], and gold-Ag nanoparticles [40]. However, complex Ag structures possess larger surface area than single Ag nanoparticles and easily aggregate. One main method to enhance the stabilization of Ag nanoparticles is to use polymers or surfactants to modify the surface of Ag particles to prevent particles from aggregating. Being a nonionic polymer compounds, polyvinyl pyrrolidone (PVP) is often used as the capping agent to control the size and shape of the colloidal nanoparticles including Ag-NPs, Au-NPs, and Pt-NPs during the particle formation [41–43]. PVP has also been reported to be a reducing agent in the preparation for the hydroxyl end-group of the PVP polymer chain [44].

In this study, we synthesized ball-like Ag nanorod aggregates via a simple seed-mediated method without any surfactant and polymeric compound as a capping reagent in reduction. PVP was added in the last step and was used only as a stabilizer. The present approach is simple, economic and green. The SERS properties of these Ag nanorod aggregates were examined by using Rhodamine 6G (R6G) and doxorubicin (DOX) as probe molecules. In addition, its catalytic performance for the reduction of 4-nitrophenol (4-NP) in the presence of $NaBH_4$ was also examined.

2. Results and Discussion

2.1. Phase Characterization

Ball-like Ag nanorod aggregates were synthesized via a seed-mediated method involving two reaction steps without using any surfactant and polymeric compound as a capping reagent in reduction. After adding 25 mL of 20 mM $AgNO_3$, aggregates comprising dozens of Ag nanorods with a mean size of about 180 nm were formed (Figure 1). It is observed that these nanorods exhibit lengths of ~50 nm and diameters of ~20 nm (Figure 1b). The polycrystalline SEAD (selected area electron diffraction) pattern of in Figure 2b confirms the diverse orientations of these nanorods in aggregates. Figure 2b shows the HRTEM image of clear lattice fringes with the spacing of 0.235 nm, which corresponds to the (111) lattice planes of the fcc-Ag [45].

Figure 1. Scanning electron microscope (SEM) images of the ball-like Ag nanorod aggregates.

Figure 2. Transmission electron microscopy (TEM) (**a**) and high resolution TEM (HRTEM) (**b**) images of the ball-like Ag nanorod aggregates. Inset in (**b**) is the selected area electron diffraction (SEAD) pattern of Ag nanorod aggregates.

2.2. UV-Vis Studies of Ag Nanorod Aggregates

It is well-known that the size and shape of metal nanoparticles could affect their optical properties such as surface plasmon resonance (SPR) property [46,47]. For instance, AgNPs with complex structures usually exhibit more than one peak [48,49], whereas spherical particles show only one size-dependent SPR peak [50]. As shown in Figure 3, the spectrum of Ag nanorod aggregates in aqueous solution displays two SPR bands that might indicate the information of nonspherical AgNPs. The lower wavelength band (435 nm) could be attributed to the out-of plane dipole resonance while the 693 nm peak (the high wavelength band) is in-plane dipole resonance [48]. In each nanorod-aggregate, the conduction electrons near each nanorod surface become delocalized and are shared amongst neighboring nanorods, which shifts the surface plasmon resonance to lower energies, moves the absorption peak to longer wavelengths and broadens the absorption spectrum.

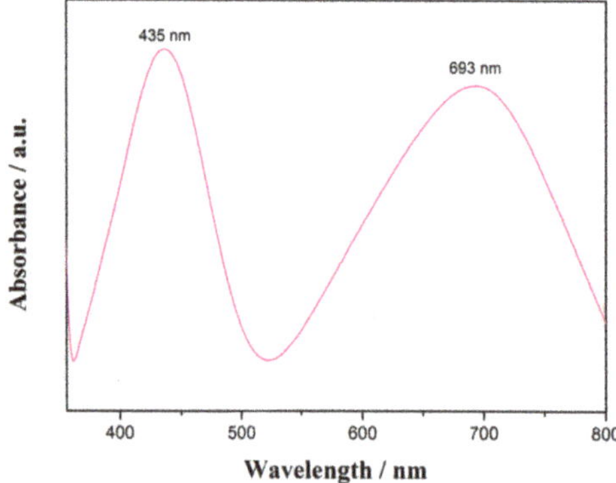

Figure 3. Ultraviolet-visible spectroscopy (UV-Vis) spectrum of the ball-like Ag nanorod aggregates.

2.3. XRD Studies of Ag Nanorod Aggregates

The structure of prepared ball-like Ag nanorod aggregates has been studied by X-ray diffraction (XRD) analysis. A typical XRD pattern of the particles was shown in Figure 4. The sharp peaks in XRD pattern prove the high crystallinity of Ag nanorod aggregates. The four diffraction peaks observed 38.17°, 44.28°, 69.45°, and 77.49° are corresponding to (111), (200), (220), and (311) Bragg's reflections of the face-centered cubic structure of Ag, respectively (JCPDS ICDD 04–0783) [51]. There is no peak of other impurities being found from the pattern, which indicates pure Ag crystals were obtained under the present method.

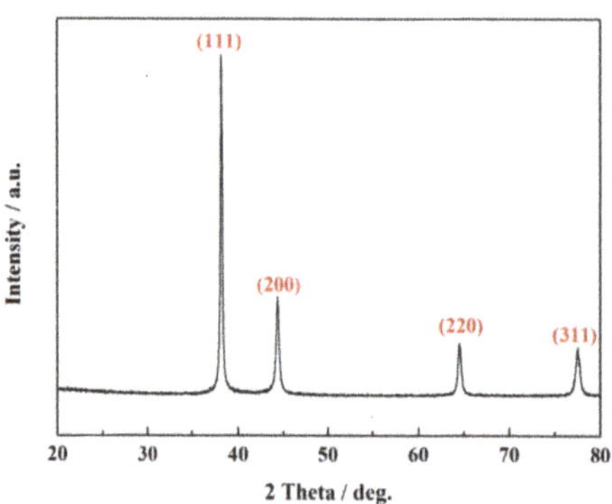

Figure 4. X-ray diffraction (XRD) pattern of the ball-like Ag nanorod aggregates.

2.4. Formation Mechanism of Ag Nanorod Aggregates

The morphology of Ag nanoparticles influences their applications. In the synthesizing process of metal nanoparticles, the morphology of nanoparticle can be controlled by adjusting the reaction time, the concentration of the precursor and the reactants, and so on [30]. In our synthesis process, the reaction was almost instantaneous. Hence, the reaction rate is not main consideration. Herein, the added Ag seeds serve as the nucleation sites for the growth of the Ag nanorod aggregates. Since the ascorbic acid used as the reducing agent in second step is excessive, the anisotropic growth process could be dominated by the amount of Ag^{+} ions, namely the concentration of $AgNO_3$. As shown in Figure 5, when the concentration of $AgNO_3$ varied from 5 mM to 20 mM, Ag nanorod aggregates show similar diameters but different morphologies. At low concentration of $AgNO_3$, a great quantity of near-spherical particles is produced. At higher concentration of $AgNO_3$, ball-like Ag nanorod aggregates are formed, indicating that the concentration of $AgNO_3$ is key factor for forming ball-like Ag nanorod aggregates [20].

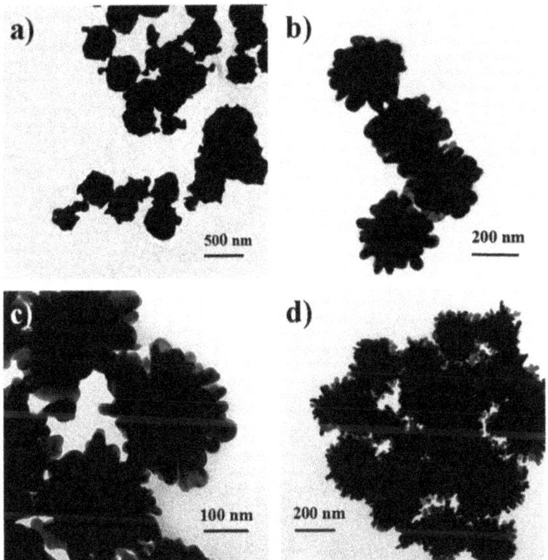

Figure 5. TEM images of the ball-like Ag nanorod aggregates under different concentrations of $AgNO_3$: (**a**) 5 mM; (**b**) 10 mM; (**c**) 15 mM; (**d**) 20 mM.

2.5. Stability Analysis of Ag Nanorod Aggregates

The aggregation of ball-like Ag nanorod aggregates is a concern for application that may take more time to handle with. To solve this problem, 0.005% wt % of PVP has been used in an effort to stabilize large size Ag particles in aqueous solution. Figure 6a shows the freshly obtained Ag nanorod aggregates without adding PVP (left) and with adding PVP (right). After 10 min, as shown in Figure 6b, Ag nanorod aggregates without adding PVP began to coagulate, while Ag nanorod aggregates with adding PVP remained stable due to the interaction between the particles and carbonyl groups on polymer chains of PVP. After 30 min (Figure 6c), Ag nanorod aggregates without adding PVP precipitated to the bottom of the bottle, and Ag nanorod aggregates with PVP begin to precipitate. However, as is shown in Figure 6d, Ag nanorod aggregates with PVP remained relatively stable even after 80 min compared with Ag nanorod aggregates without PVP. Therefore, PVP can serve as an effective stabilizer.

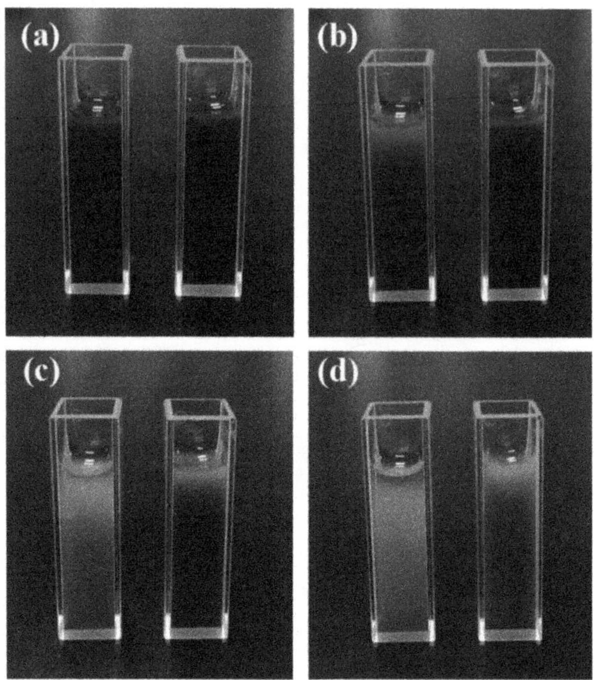

Figure 6. The coagulation condition of Ag nanorod aggregates without adding polyvinyl pyrrolidone (PVP) (left) and with adding PVP (right): (**a**) 0 min; (**b**) 10 min; (**c**) 30 min; (**d**) 80 min.

2.6. SERS Performances of Ag Nanorod Aggregates

It is critical to determine the practical limit of detection (LOD) of probe molecules in SERS applications. Accordingly, the practical LOD of R6G absorbed on ball-like Ag nanorod aggregates coated with PVP in this work was discussed. Ag nanorod aggregates formed by self-assembled nanorods. The gaps between nanorods generate active sites or hot-spots to amplify Raman scattering of probe molecules. The Raman spectra of R6G with different concentrations absorbed on AgNPs were displayed in Figure 7. All peaks of R6G in spectra agree well with previous report [52]. PVP does not produce Raman signal at such a low concentration. The peaks at 1364, 1510 and 1650 cm^{-1} are attributed to the aromatic C–C stretching modes of R6G molecules, while the peak at 772 cm^{-1} is assigned to the C–H out-of-plane bend mode. As shown by spectrum d (5×10^{-9} M), the characteristic bands of R6G at 570, 614, 1311, 1364, 1510, 1650 cm^{-1} can be still clearly detected. Therefore, the LOD for R6G absorbed on Ag nanorod aggregates was identified as 5×10^{-9} M. It is difficult to calculate the enhancement factor of the R6G molecule under available experimental conditions. Hence, we calculate the relative enhancement factor for peak at 1510 cm^{-1} by calculating the Raman intensity ratios between 5×10^{-6} M and 5×10^{-10} M. The relative enhancement factor is calculated to be 3.3×10^{3}, indicating that the flower-like nanorod aggregates could serve as effective SERS substrate.

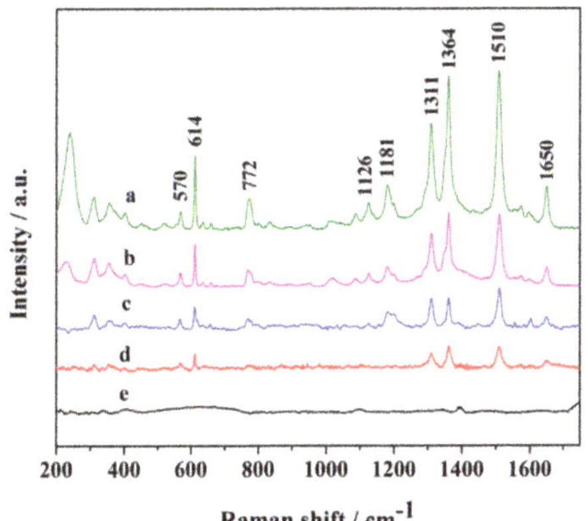

Figure 7. Raman spectra of Rhodamine 6G (R6G) at different concentrations absorbed on Ag nanorod aggregates. Spectra represent the concentrations of R6G being (**a**) 5×10^{-6}; (**b**) 5×10^{-7}; (**c**) 5×10^{-8}; (**d**) 5×10^{-9}; (**e**) 5×10^{-10} M, respectively.

Doxorubicin is commonly used as chemotherapy drug for patients with advanced cancers. SERS has been used as a powerful tool to study DOX complexes with DNA [53] and its affinity for ferric ions. In this study, we also used DOX as probe molecule to test SERS effect of Ag nanorod aggregates. Figure 8 shows the SERS spectra of DOX at different concentrations. The band at 1639 cm^{-1} is assigned to the stretching mode of carbonyl groups [54]. The band at 1296 cm^{-1} is from C–O stretching and the two strong bands at 1244 and 1210 cm^{-1} can be assigned to in-plane bending motions from C–O. The weak bands at 1082 and 795 cm^{-1} are assigned to skeletal deformations, while 990 cm^{-1} is owing to ring breath modes. When the concentration of DOX reduces to 5×10^{-6} M, bands at 1082, 1210, 1244, 1412, 1435, 1456 and 1639 cm^{-1} can still be clearly detected. Thus, the LOD for DOX absorbed on Ag nanorod aggregates was identified as 5×10^{-6} M. Hence, the flower-like Ag nanorod aggregates could serve as SERS substrate for trace analysis for small drug molecules.

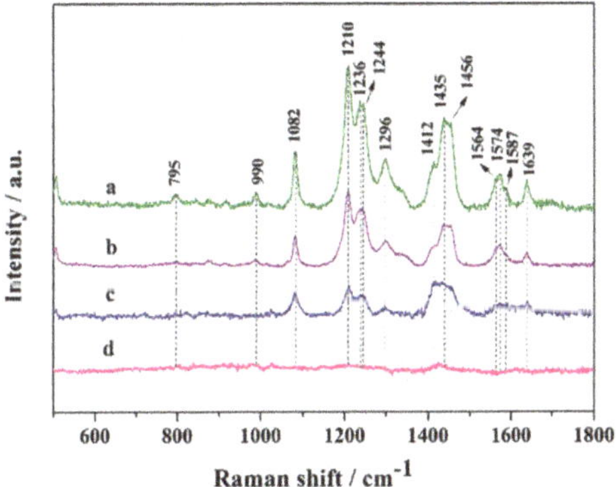

Figure 8. Raman spectra of DOX at different concentrations absorbed on Ag nanorod aggregates. Spectra represent the concentrations of DOX being (**a**) 5×10^{-4}; (**b**) 5×10^{-5}; (**c**) 5×10^{-6}; (**d**) 5×10^{-7} M, respectively.

2.7. Catalytic Reduction of 4-Nitrophenol

The reduction of 4-nitrophenol to 4-aminophenol (4-AP) by $NaBH_4$ was taken as a model reaction to examine the catalytic activity of the ball-like Ag nanorod aggregates. It is well known that the absorption peak of 4-NP with light yellow color is around 317 nm [55]. After the addition of freshly prepared $NaBH_4$ solution, the light yellow turned to intense yellow, which indicates the formation of 4-nitrophenolate ion and the pH change from acid to basic by adding $NaBH_4$. The catalytic process of this reaction was monitored by UV-Vis spectroscopy, Figure 9a shows the UV-Vis spectra of 4-NP reduction in the presence of $NaBH_4$ and 0.2 mL of Ag nanorod aggregates. As shown in Figure 8a, absorption band at 400 nm is characteristic peak of the 4-nitrophenolate in the presence of only $NaBH_4$. However, after adding 0.2 mL of ball-like Ag nanorod aggregates as a catalyst, a new band at around 300 nm emerged, indicating reduction of 4-NP to 4-AP by $NaBH_4$ (Figure 9a). The intensity of the absorption peak at 400 nm gradually decreased with time, while absorption peak at 300 nm increased simultaneously (Figure 9a). Until the intensities of two peaks no longer changed, the reduction finished. The extinction of solution at 400 nm as the function of time was measured to monitor the kinetic process of the reduction. The rate constant (K) was contingent upon reduction time and the linear plot of ln (A_t/A_0), following pseudo-first-order kinetics (Figure 9b). The constant was calculated to be 0.02252 s^{-1}, proving that the ball-like Ag aggregates is effective catalyst for the reduction of 4-NP. Usually, the catalytic activity is influenced by the surface area and roughness of the catalyst. Obviously, the good catalytic performance of the ball-like Ag nanorod aggregates could be attributed to their high surface area to volume ratio.

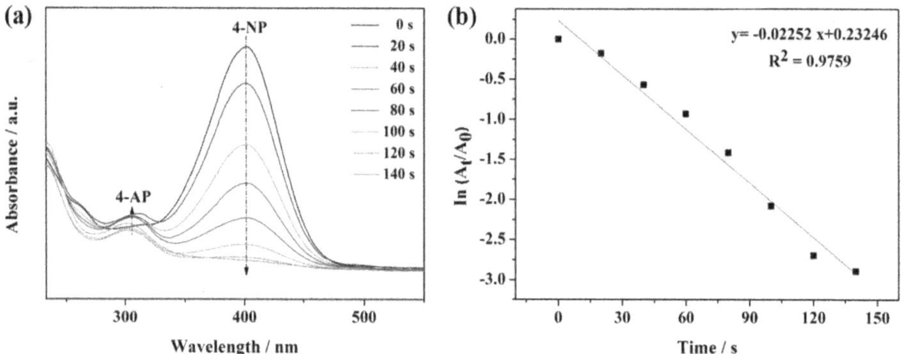

Figure 9. UV-Vis absorption spectra: (**a**) reduction of 4-NP by $NaBH_4$ using Ag nanorod aggregates as catalyst; (**b**) The plot of ln(A_t/A_0) against the reaction time for pseudo-first-order reduction kinetics of 4-NP in the presence of ball-like Ag nanorod aggregates.

3. Materials and methods

3.1. Materials

Silver nitrate ($AgNO_3$, 99.8%), tri-sodium citrate dihydrate ($C_6H_5Na_3O_7 \cdot 2H_2O$, 99%), L-ascorbic acid (Vitamin C, 99.7%), polyvinyl pyrrolidone (PVP, MW $\approx$ 45,000 daltons), Rhodamine 6G, Doxorubicin hydrochloride (DOX· HCl) was obtained from Fortuneibo-tech Co., Ltd (Shanghai, China). 4-nitrophenol (4-NP), sodium borohydride ($NaBH_4$, 96%) were purchased from Sinopharm Chemical Reagent Co. Ltd (Shanghai, China). All the chemicals were of analytical reagent grade and were used without further purification. All of the solutions were freshly prepared using deionized double-distilled water from a Milli-Q water purification system (Millipore Corporation, Billerica, MA, USA).

3.2. Preparation of Ag Aggregates

In a typical experiment, 200 mL of 2.5 mM aqueous solution of $AgNO_3$ were brought to boiling. Then, an aqueous solution of 100 mM sodium citrate (10 mL) was added dropwise in to the boiled $AgNO_3$ solution at a rate of 30 drops per min. After retained boiling for 10 min, the colorless solution turned to yellowish and turbid colloid characteristic of the seeds formation. In order to get Ag nanorod aggregates, 5 mL of Ag seed was diluted into 35 mL freshly prepared L-ascorbic acid (Vitamim C) solution (5 mM). Subsequently, 25 mL of $AgNO_3$ at a certain concentration was directly added. A generation of dark grey suspension indicated the formation of Ag nanorod aggregates. All the reactions were kept in the dark to avoid any photoreaction.

Later, 0.003 g of PVP was added to generated Ag suspension and the above mixture was performed by 5 min ultrasonic treatment. The coagulation condition of Ag particles was recorded by taking photos every 10 min in a while.

3.3. *Characterization Techniques*

Scanning electron microscopy (SEM, S-4800 II, Hitachi, Tokyo, Japan) and transmission electron microscopy (TEM, Philips Tecnai 12, Amsterdam, the Netherlands) was used to observe the morphologies and particle sizes of Ag nanorod aggregates. Transmission electron microscopy was performed by fixation on a 200-mesh carbon-coated copper grid. Ag nanorod aggregates were dropped onto clear glass slide and the glass slide was dried at room temperature. Then, the absorbance spectrum of Ag nanorod aggregates was measured by a UV-Vis spectroscopy (Varian Cary 500, Palo Alto, CA, USA) in the range of 350 to 800 nm. X-ray diffraction (XRD, D8 ADVANCE, Bruker, Karlsruhe, Germany) with graphite monochromatized Cu Kα radiation operating at 40 kV and 40 mA at room temperature in the range 2θ ($20^\circ \leqslant 2\theta \leqslant 80^\circ$) was utilized to determine the crystalline structure of the samples.

3.4. *SERS Performance of R6G and DOX on Ag Nanorod Aggregates*

To determine the LOD for R6G and DOX, a series of concentrations of R6G and DOX in water were detected using SERS Ag nanorod aggregates coated with PVP. SERS spectra were recorded with a Confocal Raman spectrometer (DXR, GX-PT-2412, Thermo, Waltham, MA, USA) with 780 nm line of a He-Ne laser as excitation wavelength. The laser power at the samples was 24 mW and the data acquisition time was 60 s. R6G and DOX was used as probe molecules. 200 μL of R6G at concentration of 5×10^{-6}, 5×10^{-7}, 5×10^{-8}, 5×10^{-9}, 5×10^{-10} M were mixed with 200 μL of Ag nanorod aggregates suspension at concentration of 8.3 mM. 200 μL of DOX at concentration of 5×10^{-4}, 5×10^{-5}, 5×10^{-6}, 5×10^{-7} M were mixed with 200 μL of Ag nanorod aggregates suspension at concentration of 8.3 mM. The spectra were obtained in solution-phase after mixing for an hour to make sure that dye molecules could absorb on the surface of AgNPs sufficiently at room temperature.

3.5. *Catalytic Reduction*

In a typical run for the reduction of 4-NP by $NaBH_4$, 0.05 mL of fresh solution of 4-NP (1 mM) was introduced into 2 mL of $NaBH_4$ (0.1 M) solution. Then, 0.2 mL of Ag nanorod aggregates (1 mM) was added to the above mixed solution. Then, with the addition of 2.75 mL of ultrapure water, the total volume of the reaction system was 5.00 mL. Set a blank group, the reaction system is same as the above system except that the Ag nanorod aggregates solution was replaced by ultrapure water. After the addition of 0.2 mL of Ag nanorod aggregates catalyst, scanning 600–250 nm band immediately in order to monitor the spectra of the 4-NP reduction in the presence of $NaBH_4$ and Ag nanorod aggregates solution by using UV-visible spectrophotometric monitoring instrument.

4. Conclusions

In summary, ball-like Ag nanorod aggregates with a mean size of 180 nm were synthesized by a seed-mediated approach, which is simple, economic and green. The polyvinyl pyrrolidone (PVP) was used to enhance the stability of obtained Ag nanorod aggregates in aqueous solution for SERS and catalytic experiment. Ag nanorod aggregates exhibit effective and reproducible SERS effect, evaluated by R6G as probe molecules. The limit of detection (LOD) for R6G and DOX are as low as 5×10^{-9} M and 5×10^{-6} M, respectively, which shows promising application for trace detection of small molecules. These Ag nanorod aggregates possess large surface area and porous surface morphology and could serve as a potential catalyst for the reduction of 4-NP in the presence of $NaBH_4$.

Acknowledgments: This work was supported by Jiangsu Province for specially appointed professorship to Peizhi Zhu, research funds from Yangzhou University, research funds from Liuda Rencai Gaofeng, the Technology Support Program of Science and Technology Department of Jiangsu Province (BE2015703), the Jiangsu agricultural science and Technology Innovation Fund Project (CX(14)2127) , the support from the Testing Center of Yangzhou University, and A Project Funded by the Priority Academic Program Development of Jiangsu Higher Education Institutions.

Author Contributions: Peizhi Zhu proposed the topic of this study and designed the experiments. Wenjing Zhang, Yin Cai and Rui Qian performed the synthesis, characterization, and analysis of AgNPs and did study on its catalytic activities. Bo Zhao performed and analyzed the SERS effects and drafted the manuscript. Peizhi Zhu and Wenjing Zhang analyzed the data and wrote the final manuscript. All authors read and approved the final manuscript.

Conflicts of Interest: The authors declare no conflict of interest.

References

1. Dasa, S.; Dhar, B.B. Green synthesis of noble metal nanoparticles using cysteine-modified silk fibroin: Catalysis and antibacterial activity. *RSC Adv.* **2014**, *4*, 46285–46292. [CrossRef]
2. Mi, S.N.; Jun, B.H.; Kim, S.; Kang, H.; Woo, M.A.; Minai-Tehrani, A.; Kim, J.E.; Kim, J.; Park, J.; Lim, H.T.; *et al.* Magnetic surface-enhanced Raman spectroscopic (M-SERS) dots for the identification of bronchioalveolar stem cells in normal and lung cancer mice. *Biomaterials* **2009**, *30*, 3915–3925.
3. Reithofer, M.R.; Lakshmanan, A.; Ping, A.T.K.; Jia, M.C.; Hauser, C.A.E. *In situ* synthesis of size-controlled, stable silver nanoparticles within ultrashort peptide hydrogels and their anti-bacterial properties. *Biomaterials* **2014**, *35*, 7535–7542. [CrossRef] [PubMed]
4. Wang, Y.L.; Lee, K.; Irudayaraj, J. SERS aptasensor from nanorod-nanoparticle junction for protein detection. *Chem. Commun.* **2010**, *46*, 613–615. [CrossRef] [PubMed]
5. Shafer-Peltier, K.E.; Haynes, C.L.; Glucksberg, M.R.; van Duyne, R.P. Toward a glucose biosensor based on surface-enhanced Raman scattering. *J. Am. Chem. Soc.* **2003**, *125*, 588–593. [CrossRef] [PubMed]
6. Hou, M.J.; Huang, Y.; Ma, L.W.; Zhang, Z.J. Sensitivity and reusability of SiO_2 NRs@AuNPs SERS substrate in trace monochlorobiphenyl detection. *Nanoscale Res. Lett.* **2015**, *10*. [CrossRef] [PubMed]
7. Li, J.M.; Ma, W.F.; Wei, C.; You, L.J.; Guo, J.; Hu, J.; Wang, C.C. Detecting trace melamine in solution by SERS using Ag nanoparticle coated poly(styrene-co-acrylic acid) nanospheres as novel active substrates. *Langmuir* **2011**, *27*, 14539–14544. [CrossRef] [PubMed]
8. Li, J.F.; Huang, Y.F.; Ding, Y.; Yang, Z.L.; Li, S.B.; Zhou, X.S.; Fan, F.R.; Zhang, W.; Zhou, Z.Y.; Wu, Y.D.; *et al.* Shell-isolated nanoparticle-enhanced Raman spectroscopy. *Nature* **2010**, *464*, 392–395. [CrossRef] [PubMed]
9. Cheng, M.L.; Tsai, B.C.; Yang, J. Silver nanoparticle-treated filter paper as a highly sensitive surface-enhanced Raman scattering (SERS) substrate for detection of tyrosine in aqueous solution. *Anal. Chim. Acta* **2011**, *708*, 89–96. [CrossRef] [PubMed]
10. Maiti, K.K.; Dinish, U.S.; Samanta, A.; Vendrell, M.; Soh, K.S.; Park, S.J.; Olivo, M.; Chang, Y.T. Multiplex targeted *in vivo* cancer detection using sensitive near-infrared SERS nanotags. *Nano Today* **2012**, *7*, 85–93. [CrossRef]
11. Herrera, G.M.; Padilla, A.C.; Hernandez-Rivera, S.P. Surface enhanced Raman scattering (SERS) studies of gold and silver nanoparticles prepared by laser ablation. *Nanomaterials* **2013**, *3*, 158–172. [CrossRef]
12. Li, Y.S.; Cheng, Y.Y.; Xu, L.P.; Du, H.W.; Zhang, P.X.; Wen, Y.Q.; Zhang, X.J. A nanostructured SERS switch based on molecular beacon-controlled assembly of gold nanoparticles. *Nanomaterials* **2016**, *6*. [CrossRef]
13. Onuegbu, J.; Fu, A.; Glembocki, O.; Pokes, S.; Alexson, D.; Hosten, C.M. Investigation of chemically modified barium titanate beads as surface-enhanced Raman scattering (SERS) active substrates for the detection of benzene thiol, 1,2-benzene dithiol, and rhodamine 6G. *Spectrochim. Acta A* **2011**, *79*. [CrossRef] [PubMed]
14. Fateixa, S.; Pinheiro, P.C.; Nogueira, H.I.S.; Trindade, T. Composite blends of gold nanorods and poly(*t*-butylacrylate) beads as new substrates for SERS. *Spectrochim. Acta A* **2013**, *113*. [CrossRef] [PubMed]
15. Li, S.Q.; Liu, L.; Hu, J.B. An approach for fabricating self-assembled monolayer of gold nanoparticles on NH^{2+} ion implantation modified indium tin oxide as the SERS-active substrate. *Spectrochim. Acta A* **2012**, *86*, 533–537. [CrossRef] [PubMed]

16. Philip, D.; Gopchandran, K.G.; Unni, C.; Nissamudeen, K.M. Synthesis, characterization and SERS activity of Au-Ag nanorods. *Spectrochim. Acta A* **2008**, *70*, 780–784. [CrossRef] [PubMed]
17. Yang, L.B.; Qin, X.Y.; Gong, M.D.; Jiang, X.; Yang, M.; Li, X.L.; Li, G.Z. Improving surface-enhanced Raman scattering properties of TiO_2 nanoparticles by metal Co doping. Spectrochim. *Acta A* **2014**, *123*, 224–229. [CrossRef] [PubMed]
18. Wu, W.J.; Wu, M.Z.; Sun, Z.Q.; Li, G.; Ma, Y.Q.; Liu, X.S.; Wang, X.F.; Chen, X.S. Morphology controllable synthesis of silver nanoparticles: Optical properties study and SERS application. *J. Alloy.Compd.* **2013**, *579*, 117–123. [CrossRef]
19. Sun, L.L.; Song, Y.H.; Wang, L.; Guo, C.L.; Sun, Y.J.; Liu, Z.L.; Li, Z. Ethanol-induced formation of silver nanoparticle aggregates for highly active SERS substrates and application in DNA detection. *J. Phys. Chem. C* **2008**, *112*, 1415–1422. [CrossRef]
20. Nhung, T.T.; Lee, S.W. Green synthesis of asymmetrically textured silver meso-flowers (AgMFs) as highly sensitive SERS substrates. *ACS Appl. Mater. Interfaces* **2014**, *6*, 21335–21345. [CrossRef] [PubMed]
21. Zhang, M.F.; Zhao, A.W.; Sun, H.H.; Guo, H.Y.; Wang, D.P.; Li, D.; Gan, Z.B.; Tao, W.Y. Rapid, large-scale, sonochemical synthesis of 3D nanotextured silver microflowers as highly efficient SERS substrates. *J. Mater. Chem.* **2011**, *21*, 18817–18824. [CrossRef]
22. Dao, A.T.N.; Mott, D.M.; Higashimine, K.; Maenosono, S. Enhanced electronic properties of Pt@Ag heterostructured nanoparticles. *Sensors* **2013**, *13*, 7813–7826. [CrossRef] [PubMed]
23. Qayyum, E.; Castillo, V.A.; Warrington, K.; Barakat, M.A.; Kuhn, J.N. Methanol oxidation over silica-supported Pt and Ag nanoparticles: Toward selective production of hydrogen and carbon dioxide. *Catal. Commum.* **2012**, *28*, 128–133. [CrossRef]
24. Lippits, M.J.; Nieuwenhuys, B.E. Direct conversion of ethanol into ethylene oxide on copper and silver nanoparticles: Effect of addition of CeO_x and Li_2O. *Catal. Today* **2010**, *154*, 127–132. [CrossRef]
25. Wunder, S.; Polzer, F.; Lu, Y.; Mei, Y.; Ballauff, M. Kinetic analysis of catalytic reduction of 4-nitrophenol by metallic nanoparticles immobilized in spherical polyelectrolyte brushes. *J. Phys. Chem. C* **2010**, *114*, 8814–8820. [CrossRef]
26. Yang, J.H.; Cao, B.B.; Li, H.Q.; Liu, B. Investigation of the catalysis and SERS properties of flower-like and hierarchical silver microcrystals. *J. Nanopart. Res.* **2014**, *16*. [CrossRef]
27. Sajanlal, P.R.; Pradeep, T. Mesoflowers: A new class of highly efficient surface-enhanced Raman active and infrared-absorbing materials. *Nano. Res.* **2009**, *2*, 306–320. [CrossRef]
28. Xia, J.R.; Wei, R.; Wu, Y.M.; Li, W.H.; Yang, L.N.; Yang, D.H.; Song, P. Synthesis of large flower-like substrates for surface-enhanced Raman scattering. *Chem. Eng. J.* **2014**, *244*, 252–257. [CrossRef]
29. Kar, S.; Desmonda, C.; Tai, Y. Synthesis of SERS-active stable anisotropic silver nanostructures constituted by self-assembly of multiple silver nanopetals. *Plasmonics* **2014**, *9*, 485–492. [CrossRef]
30. Zhou, N.; Li, D.S.; Yang, D.R. Morphology and composition controlled synthesis of flower-like silver nanostructures. *Nanoscale Res. Lett.* **2014**, *9*. [CrossRef] [PubMed]
31. Xu, M.W.; Zhang, Y. Seed-mediated approach for the size-controlled synthesis of flower-like Ag mesostructures. *Mater. Lett.* **2014**, *130*, 9–13. [CrossRef]
32. Tang, B.; Xu, S.P.; Jian, X.G.; Tao, J.L.; Xu, W.Q. Real-time, in-situ, extinction spectroscopy studies on silver-nanoseed formation. *Appl. Spectrosc.* **2010**, *64*, 1407–1415. [CrossRef] [PubMed]
33. Mahl, D.; Diendirf, J.; Ristig, S.; Greulich, C.; Li, Z.A.; Farle, M.; Köller, M.; Epple, M. Silver, gold, and alloyed silver-gold nanoparticles: Characterization and comparative cell-biologic action. *J. Nanopart. Res.* **2012**, *14*. [CrossRef]
34. Noh, J.H.; Meijboom, R. Catalytic evaluation of dendrimer-templated Pd nanoparticles in the reduction of 4-nitrophenol using Langmuir-Hinshelwood kinetics. *Appl. Surf. Sci.* **2014**, *320*, 400–413. [CrossRef]
35. Gutés, A.; Carraro, C.; Maboudian, R. Silver dendrites from galvanic displacement on commercial aluminum foil as an effective SERS substrate. *J. Am. Chem. Soc.* **2010**, *132*, 1476–1477. [CrossRef] [PubMed]
36. Zhang, Q.; Li, W.Y.; Moran, C.; Zeng, J.; Chen, J.Y.; Wen, L.P.; Xia, Y.N. Seed-mediated synthesis of Ag nanocubes with controllable edge lengths in the range of 30–200 nm and comparison of their optical properties. *J. Am. Chem. Soc.* **2010**, *132*, 11372–11378. [CrossRef] [PubMed]
37. Zhang, L.; Wang, Y.; Tong, L.M.; Xia, Y.N. Seed-mediated synthesis of silver nanocrystals with controlled sizes and shapes in droplet microreactors separated by air. *Langmuir* **2013**, *29*, 15719–15725. [CrossRef] [PubMed]

38. Guo, S.J.; Zhang, S.; Su, D.; Sun, S.H. Seed-mediated synthesis of core/shell FePtM/FePt (M = Pd, Au) nanowires and their electrocatalysis for oxygen reduction reaction. *J. Am. Chem. Soc.* **2013**, *135*, 13879–13884. [CrossRef] [PubMed]
39. Xia, X.H.; Zeng, J.; Oetjen, L.K.; Li, Q.G.; Xia, Y.N. Quantitative analysis of the role played by poly(vinylpyrrolidone) in seed-mediated growth of Ag nanocrystals. *J. Am. Chem. Soc.* **2012**, *134*, 1793–1801. [CrossRef] [PubMed]
40. Mcgilvray, K.L.; Fasciani, C.; Bueno-Alejo, C.J.; Schwartz-Narbonne, R.; Scaiano, J.C. Photochemical strategies for the seed-mediated growth of gold and gold-silver nanoparticles. *Langmuir* **2012**, *28*, 16148–16155. [CrossRef] [PubMed]
41. Liang, H.Y.; Wang, W.Z.; Huang, Y.Z.; Zhang, S.P.; Wei, H.; Xu, H.X. Controlled synthesis of uniform silver nanospheres. *J. Phys. Chem. C* **2010**, *114*, 7427–7431. [CrossRef]
42. Tang, X.L.; Jiang, P.; Ge, G.L.; Tsuji, M.; Xie, S.S.; Guo, Y.J. Poly(*N*-vinyl-2-pyrrolidone) (PVP)-capped dendritic gold nanoparticles by a one-step hydrothermal route and their high SERS effect. *Langmuir* **2008**, *24*, 1763–1768. [CrossRef] [PubMed]
43. Hossain, M.J.; Tsunoyama, H.; Yamauchi, M.; Ichikuni, N.; Tsukuda, T. High-yield synthesis of PVP-stabilized small Pt clusters by microfluidic method. *Catal. Today* **2012**, *183*, 101–107. [CrossRef]
44. Wu, C.W.; Mosher, B.P.; Lyons, K.; Zeng, T.F. Reducing ability and mechanism for polyvinylpyrrolidone (PVP) in silver nanoparticles synthesis. *J. Nanosci. Nanotechnol.* **2010**, *10*, 2342–2347. [CrossRef] [PubMed]
45. Wang, A.L.; Yin, H.B.; Ren, M.; Liu, Y.M.; Jiang, T.S. Synergistic effect of silver seeds and organic modifiers on the morphology evolution mechanism of silver nanoparticles. *Appl. Surf. Sci.* **2008**, *254*, 6527–6536. [CrossRef]
46. Rashid, M.H.; Mandal, T.K. Synthesis and catalytic application of nanostructured silver dendrites. *J. Phys. Chem. C* **2007**, *111*, 16750–16760. [CrossRef]
47. Mayer, K.M.; Hafner, J.H. Localized surface plasmon resonance sensors. *Chem. Rev.* **2011**, *111*, 3828–3857. [CrossRef] [PubMed]
48. Chen, S.H.; Carroll, D.L. Synthesis and characterization of truncated triangular silver nanoplates. *Nano Lett.* **2002**, *2*, 1003–1007. [CrossRef]
49. Zou, X.Q.; Ying, E.B.; Dong, S.J. Preparation of novel silver-gold bimetallic nanostructures by seeding with silver nanoplates and application in surface-enhanced Raman scattering. *J. Colloid Interf. Sci.* **2007**, *306*, 307–315. [CrossRef] [PubMed]
50. Pillai, Z.S.; Kamat, P.V. What factors control the size and shape of silver nanoparticles in the citrate ion reduction method? *J. Phys. Chem. B* **2004**, *108*, 945–951. [CrossRef]
51. Ahmad, M.B.; Lim, J.J.; Shameli, K.; Ibrahim, N.A.; Tay, M.Y. Synthesis of silver nanoparticles in chitosan, gelatin and chitosan/gelatin bionanocomposites by a chemical reducing agent and their characterization. *Molecules* **2011**, *16*, 7237–7248. [CrossRef] [PubMed]
52. Li, Y.X.; Zhang, K.; Zhao, J.J.; Ji, J.; Ji, C.; Liu, B.H. A three-dimensional silver nanoparticles decorated plasmonic paper strip for SERS detection of low-abundance molecules. *Talanta* **2016**, *147*, 493–500. [CrossRef] [PubMed]
53. Beljebbar, A.; Sockalingum, G.D.; Angiboust, J.F.; Manfait, M. Comparative FT SERS, resonance Raman and SERRS studies of doxorubicin and its complex with DNA. *Spectrochim. Acta A* **1995**, *51*, 2083–2090. [CrossRef]
54. Gautier, J.; Munnier, E.; Douziech-Eyrolles, L.; Paillard, A.; Dubois, P.; Chourpa, L. SERS spectroscopic approach to study doxorubicin complexes with Fe^{2+} ions and drug release from SPION-based nanocarriers. *Analyst* **2013**, *138*, 7354–7361. [CrossRef] [PubMed]
55. Baruah, B.; Gabriel, G.J.; Akbashev, M.J.; Booher, M.E. Facile synthesis of silver nanoparticles stabilized by cationic polynorbornenes and their catalytic activity in 4-nitrophenol reduction. *Langmuir* **2013**, *29*, 4225–4234. [CrossRef] [PubMed]

Article

Enhanced Activity of Supported Ni Catalysts Promoted by Pt for Rapid Reduction of Aromatic Nitro Compounds

Huishan Shang [1,2], Kecheng Pan [1], Lu Zhang [1], Bing Zhang [2,*] and Xu Xiang [1,*]

1 State Key Laboratory of Chemical Resource Engineering, Beijing University of Chemical Technology, Beijing 100029, China; huishan6880220@163.com (H.S.); xiangxubit@sohu.com (K.P.); zhanglu@mail.buct.edu.cn (L.Z.)

2 School of Chemical Engineering, Zhengzhou University, Zhengzhou 450001, China

* Correspondence: zhangb@zzu.edu.cn (B.Z.); xiangxu@mail.buct.edu.cn (X.X.); Tel.: +86-371-6778-1724 (B.Z.); +86-10-6443-2931 (X.X.)

Academic Editors: Hermenegildo García and Sergio Navalón
Received: 22 April 2016; Accepted: 13 May 2016; Published: 4 June 2016

Abstract: To improve the activities of non-noble metal catalysts is highly desirable and valuable to the reduced use of noble metal resources. In this work, the supported nickel (Ni) and nickel-platinum (NiPt) nanocatalysts were derived from a layered double hydroxide/carbon composite precursor. The catalysts were characterized and the role of Pt was analysed using X-ray diffraction (XRD), high-resolution transmission electron microscopy (HRTEM), energy dispersive X-ray spectroscopy (EDS) mapping, and X-ray photoelectron spectroscopy (XPS) techniques. The Ni^{2+} was reduced to metallic Ni^0 via a self-reduction way utilizing the carbon as a reducing agent. The average sizes of the Ni particles in the NiPt catalysts were smaller than that in the supported Ni catalyst. The electronic structure of Ni was affected by the incorporation of Pt. The optimal NiPt catalysts exhibited remarkably improved activity toward the reduction of nitrophenol, which has an apparent rate constant (K_a) of 18.82×10^{-3} s^{-1}, 6.2 times larger than that of Ni catalyst and also larger than most of the reported values of noble-metal and bimetallic catalysts. The enhanced activity could be ascribed to the modification to the electronic structure of Ni by Pt and the effect of exposed crystal planes.

Keywords: supported catalysts; heterogeneous catalysis; nickel; platinum; reduction

1. Introduction

Aromatic nitro compounds are widely generated as byproducts in various industries, including in the production of pigments, pesticides and medicines [1]. 4-nitrophenol (4-NP) is among the most common aromatic nitro compounds, and is harmful to the environment [2]. 4-aminophenol (4-AP), the reduction product of 4-NP, is an important intermediate for the manufacture of dyes, agrochemicals, and pharmaceuticals [3–5].

Various methods to synthesize 4-AP have been reported, such as multi-step iron-acid reduction of 4-NP, catalytic reduction of nitrobenzene, and electrochemical synthesis [6–8]. Among these methods, catalytic reduction is an alternative green process for 4-AP production because it does not generate a large amount of un-reusable Fe–FeO sludges and acid/alkali effluents [9]. It is well known that diverse noble metal such as palladium, platinum, and gold catalysts have been widely used in the catalytic reactions due to their high catalytic activities [10–13]. However, the high cost and scarcity in nature limit their practical applications. Therefore, non-noble metal catalysts have been paid more attention because of their abundance and reduced cost [14]. As we know, the reduction rates of 4-NP to 4-AP over non-noble metal catalysts are far slower than noble metal ones [15]. Therefore, it is desirable to improve the catalytic activities of the supported non-noble metal catalysts.

Among non-noble metals, nickel nanoparticles (Ni NPs) have drawn much attention because of their easy availability, relatively high catalytic activity, and magnetic separation feature [16,17]. Moreover, the unique physical and chemical properties can be introduced by adding a small amount of second elements such as Pd, Pt, Ir, or Ru to Ni catalysts. These elements are dispersed on the surface of Ni to form bimetallic catalysts and the interactions at the metal–metal interface lead to the improved performance of Ni catalysts [18–21]. Thus, a promising endeavor is to explore a combination of noble metal, e.g., Pt and Ni NPs to enhance the catalytic activity. In addition, the Ni-based catalysts can be easily separated from the reaction medium by application of an external magnetic field.

In the previous work, we prepared supported Ni catalysts via self-reduction of hybrid NiAl-layered double hydroxide/carbon (NiAl-LDH/C) composites [22]. The composites were assembled via crystallization of LDH in combination with simultaneous carbonization of glucose under hydrothermal conditions. The resulting carbon acted as a reducing agent to convert nickel oxide to metallic nickel upon calcination. The as-synthesized Ni nanoparticles had small crystallite sizes and high dispersions in the support owing to the confined effects of LDH and carbon. Inspired by this finding, we extended this method to prepare Pt-modified Ni catalysts. The glucose-derived carbon in the composites led to *in situ* reduction of chloroplatinic acid anions upon hydrothermal carbonization and thus introduced Pt to the Ni catalysts. The catalytic activities of the resultant Pt-modified Ni catalysts for liquid-phase reduction of 4-NP to 4-AP were investigated under mild reaction conditions. The optimal NiPt catalysts exhibited larger apparent rate constant (K_a) than most of the reported values of noble-metal and bimetallic catalysts.

2. Results and Discussions

2.1. Characterization of Materials

The XRD patterns of LDH, hybrid LDH/C and Pt@LDH/C composites were shown in Figure 1a. The pattern of LDH sample exhibited the characteristic (003), (006), (009), (110), and (113) reflections, corresponding to layered hydrotalcite-like compounds [23]. The intensive reflections revealed the highly crystalline nature of the product. In contrast, the patterns of LDH/C and Pt@LDH/C composites presented broadened reflections at the same 2θ positions as those of LDH. It was noted that no graphite or other forms of carbon phases were detected, suggesting that carbon products mostly existed in amorphous form. The broader and weaker reflections for LDH/C and Pt@LDH/C were caused by both the dilution effect of the resultant carbon component in the composites and the reduced crystalline feature of LDH phase. No reflections associated with metal Pt or other Pt species were observed in Pt@LDH/C. This finding could be ascribed to the high dispersion of the Pt nanoparticles of small sizes and/or the very low loading [24]. It was believed that the aromatization and carbonization of glucose yielded amorphous carbon under the present hydrothermal treatment [25], and the resultant carbon was assembled with LDH crystallites during the crystallization of LDH, thereby leading to the formation of hybrid composites.

The effects of calcination temperatures on the phase structure were also studied. Figure 1b showed XRD patterns of Pt@LDH/C precursor calcined at 500 °C, 600 °C, and 700 °C in a flowing N_2 atmosphere. It was known that the layered structure of LDH collapses and is converted into mixed-oxide phases when heating above 450 °C [26]. For NiPt-0.6% (500) and NiPt-0.6% (600) samples, the XRD patterns exhibited the co-existence of Ni and NiO phases. Furthermore, the relative intensity from the reflections of NiO decreased with increasing calcination temperature. After elevating the temperature to 700 °C, only the contributions from metallic Ni were observed, which can be indexed to the face-centered cubic (fcc) Ni phase (JCPDS No. 87-0712). The three peaks centered at 2θ = 44.4°, 51.7° and 76.5° corresponded to the characteristic (111), (200) and (220) planes of Ni phase. The results indicated that the complete phase transformation from NiO to Ni occurred at this temperature.

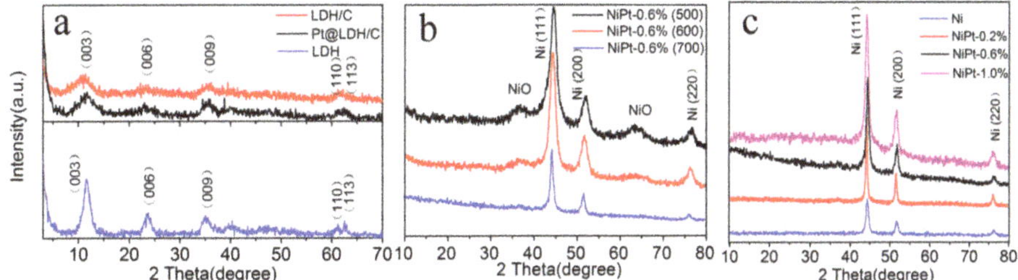

Figure 1. X-ray diffraction (XRD) patterns of (**a**) layered double hydroxide (LDH), LDH/carbon (C), and Pt@LDH/C; (**b**) NiPt catalysts obtained at different calcination temperatures; and (**c**) the Ni and NiPt catalysts with different Pt loadings.

The LDH-derived metal oxides could be reduced into metal via self-reduction by carbon in the LDH/C composite, leading to the tranformation of Ni^{2+} to Ni^0. The main reactions can be expressed as follows:

$$NiO + C\,(s) \rightarrow Ni + CO\,(g) \tag{1}$$

$$NiO + CO\,(g) \rightarrow Ni + CO_2\,(g) \tag{2}$$

During heating, LDH was transformed into mixed oxides (crystalline NiO and amorphous alumina). After reduction, the residual carbon acted not only as the support but also as the dispersing matrix, which prevented Ni particles from agglomeration. Figure 1c showed the XRD patterns of NiPt catalysts with different Pt loadings. These XRD patterns of the Ni and NiPt samples were hardly distinguishable, which could indicate that the Pt species were highly dispersed in the NiPt catalysts [27].

TEM observations were carried out to investigate the morphologies and microstructures of NiPt catalysts with different Pt loadings (Figure 2). TEM-derived histograms of the Ni particle size distributions were presented. The TEM images indicated that the Ni nanoparticles were well dispersed on the support and the size slightly decreased from 12.1 to 11.0 nm when the nominal Pt content was increased to 0.6% compared to the pristine Ni. When the Pt content was increased to 1.0%, the Ni nanoparticles had smaller size (~9 nm) and narrower size distributions (Table 1). The Ni loading reached as high as 36%–39% whereas no obvious aggregates were observed, suggestive of excellent metal dispersion. The Pt particles could not be found in the TEM images owing to the low loading. The Pt loading measured was lower than the nominal value, possibly due to the metal leaching during the hydrothermal reactions (Table 1). One can find that these catalysts had similar specific surface areas, which indicated that the surface area could not be the dominant factor to adjust the catalytic properties [28].

Table 1. Metal loadings, sizes and specific surface areas of the supported catalysts.

Catalyst	Pt wt % [1]	Ni wt %	D (nm) [2]	Specific Surface Area ($m^2\,g^{-1}$) [3]
Ni	0	35.85	12.1	244.6
NiPt-0.2%	0.167	38.75	12.1	271.8
NiPt-0.6%	0.347	36.98	11.0	266.0
NiPt-1.0%	0.537	39.55	9.0	267.4

[1] Pt content measured by inductively coupled plasma-atomic emission spectroscopy (ICP-AES); [2] The mean size of Ni nanoparticles (NPs) based on TEM analysis; [3] Specific surface area calculated by a Brunauer-Emmett-Teller (BET) method.

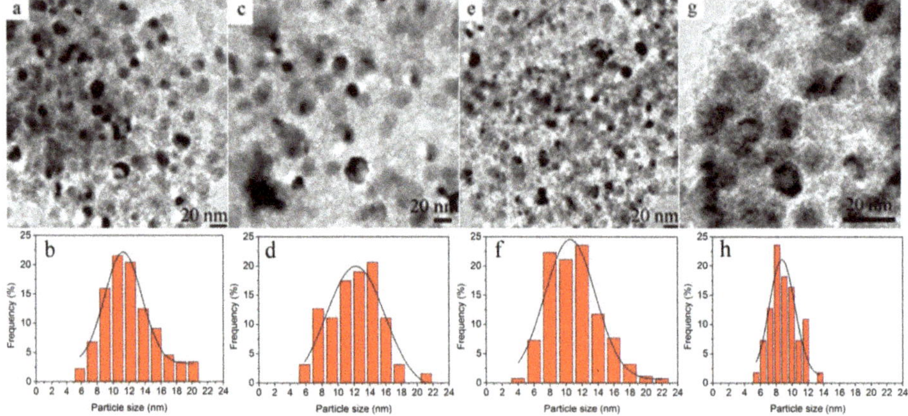

Figure 2. Transmission electron microscopy (TEM) images and size distributions of the supported Ni and NiPt catalysts with different Pt loadings (**a**) and (**b**) Ni; (**c**) and (**d**) NiPt-0.2%; (**e**) and (**f**) NiPt-0.6%; (**g**) and (**h**) NiPt-1.0%.

To further reveal the microstructures of the Ni and Pt NPs, high-resolution transmission electron microscopy (HRTEM) observations were conducted. As shown in Figure 3a, the lattice spacing of the supported Ni NPs was measured to be 0.203 nm, consistent with the d-value of plane (111) of cubic phase Ni. Regarding NiPt-0.2% and NiPt-1.0% catalysts, the Ni NPs had the same d-spacing of 0.203 nm (Figure 3b,d), and the Pt NPs exhibited a d-spacing of 0.230 nm, corresponding to plane (111) of Pt. The adjacent interface could be observed between plane Ni (111) and Pt (111). As to NiPt-0.6% catalysts, the d-spacing of 0.170 nm for Ni NPs and 0.193 nm for Pt NPs were observed, which corresponded to Ni (200) and Pt (200), respectively (Figure 3c). It was interesting that the interface formed between Ni (111) and Pt (111) or Ni (200) and Pt (200). Such interface of different planes of Ni and Pt NPs could affect the catalytic properties. It is known that the catalytic reduction usually depends on the sizes, shapes, and/or exposed planes of the active metal NPs [23,29]. The differences in exposed planes and interfaces of NiPt catalysts could lead to distinct activities towards the reduction reaction. Energy dispersive X-ray spectroscopy (EDS) mapping analyses of NiPt-0.6% exhibited C, O, Al, Ni, and Pt elements (Figure 3e). It was seen that C, O and Al were homogeneously distributed in the whole matrix. The Ni components were highly dispersed on the matrix and as well the small amount of Pt.

X-ray photoelectron spectroscopy (XPS) measurements were used to analyze the chemical states of the metal species in the catalysts. The XPS spectra of Pt and Ni core levels were shown in Figure 4. The Pt4f spectra could be deconvoluted into two peaks due to the spin-orbit splitting. One was at 70.9–71.4 eV, and the other was at 74.2–75.0 eV, which could be assigned to metallic Pt^0 species [30,31]. As to NiPt-1.0% sample, the Pt4f peaks showed a 0.5–0.8 eV shift towards a higher binding energy compared to the other two NiPt catalysts (Table 2). This was an indication that the electronic structures of Pt were affected by the surrounding Ni NPs owing to the strong interactions between them [32]. The Ni2p spectra were deconvoluted into three contributions at ~853.3, ~856.8, and ~862.5 eV in the $Ni2p_{3/2}$ region, which were assigned to the metallic Ni^0, the Ni^{2+} species in NiO and the satellite, respectively [33]. The existence of Ni^{2+} species could be due to the easy oxidation of metallic Ni, which inevitably contact with the air during the sample preparation and settlement [22]. Compared to the Ni catalysts, the binding energy of Ni^0 and Ni^{2+} ($Ni2p_{3/2}$) in the NiPt-0.2% had a little shift of 0.1–0.2 eV towards a lower value. The binding energy shifted ~0.4 eV with the increasing Pt content in the NiPt-0.6% and NiPt-1.0% samples (Table 2). The binding energy of metallic Ni shifted to a lower value after a small amount of Pt was incorporated. It is a hint that the electronic structures of Ni were affected by Pt, suggesting the strong interactions between the two metal components. The binding energy at 68.9 eV could be the contribution from Ni3p [34,35]. The Ni3p peak overlapped with the $Pt4f_{7/2}$ peak, which provided further evidence for the existence of interactions between Ni and Pt. The binding energy at 74.8 ± 0.1 eV was assigned to Al2p core level, which came from the Al_2O_3 matrix in the catalysts [36]. XPS characterization verified that the interactions between Ni and Pt were enhanced with the increasing content of Pt.

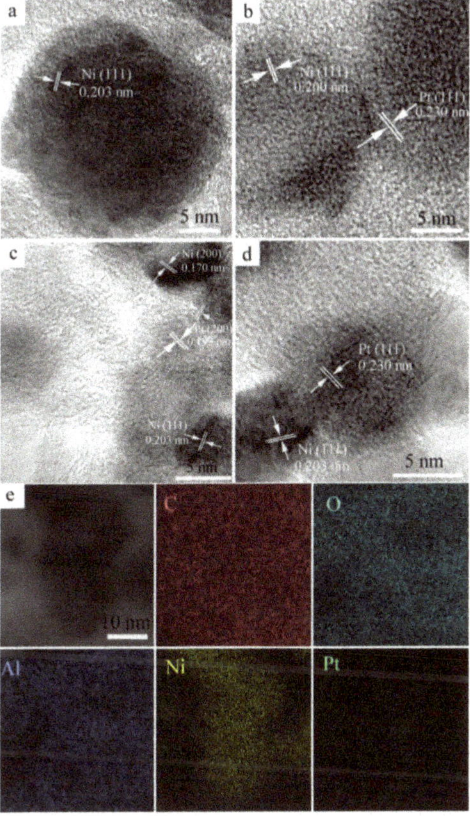

Figure 3. High-resolution transmission electron microscopy (HRTEM) images of the supported catalysts: (**a**) Ni; (**b**) NiPt-0.2%; (**c**) NiPt-0.6%; (**d**) NiPt-1.0%; (**e**) Energy dispersive X-ray spectroscopy (EDS) mapping of the catalyst NiPt-0.6%.

Table 2. XPS binding energy of the Ni, Pt and Al core levels in the supported catalysts.

Catalysts	$2p_{3/2}$ (eV)			$2p_{1/2}$ (eV)			$4f_{7/2}$ (eV)	$4f_{5/2}$ (eV)	Ni3p	Al2p
	Ni^0	Ni^{2+}	$Ni^{Sat.}$	Ni^0	Ni^{2+}	$Ni^{Sat.}$	Pt^0			
Ni	853.3	856.8	862.5	870.9	874.4	880.5	-	-	-	74.8
NiPt-0.2%	853.2	856.6	862.1	870.1	874.4	880.2	70.9	74.2	68.1	74.9
NiPt-0.6%	852.8	856.4	862.1	870.1	874.1	880.3	70.9	74.4	67.7	74.7
NiPt-1.0%	852.7	856.3	862.0	870.0	874.0	880.1	71.4	75.0	68.4	74.9

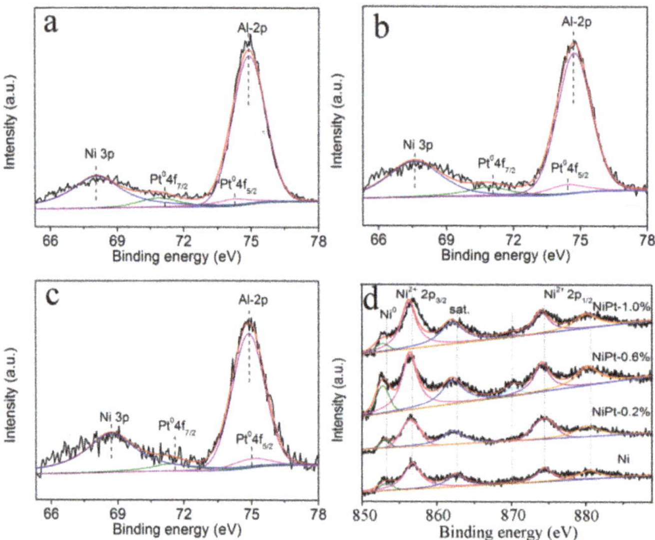

Figure 4. X-ray photoelectron spectroscopy (XPS) spectra of Pt and Ni core levels in the supported catalysts: (**a**) NiPt-0.2%; (**b**) NiPt-0.6%; (**c**) NiPt-1.0% and (**d**) Ni.

2.2. Evaluation of the Catalytic Activity

The catalytic activities of the as-prepared Ni and NiPt (0.2%, 0.6%, and 1.0%) catalysts were evaluated using a probe reaction, *i.e.*, the reduction of 4-NP into 4-AP. The reduction was conducted in an aqueous solution of 4-NP with the addition of $NaBH_4$ as a reducing agent. The Ultraviolet-visible (UV-VIS) absorption spectra were recorded with time to monitor the transformation from 4-NP to 4-AP due to their distinct absorption positions. The original adsorption of 4-NP peaked at 317 nm (Figure 5a). The absorption shifted to 400 nm when the freshly prepared aqueous solution of $NaBH_4$ was added, and the color of the solution immediately changed from the pristine light yellow to bright yellow (the inset of Figure 5a). This redshift was ascribed to the rapid formation of 4-nitrophenolate ions in alkaline solution upon the addition of $NaBH_4$ [37]. When the NiPt-0.6% catalysts were added to the solution, the bright yellow solution faded and became colorless in tens of seconds. The absorption at 400 nm sharply decreased, and the absorption at 300 nm from 4-AP appeared [38]. The evolution of absorbance spectra with time clearly showed the conversion from 4-NP to 4-AP in the presence of NiPt-0.6% catalysts (Figure 5b). The reduction was completed within around 100 seconds according to the decreasing absorbance at 400 nm and the increasing absorbance at 300 nm, corresponding to the deletion of 4-NP and the formation of 4-AP, respectively.

The Ni and NiPt catalysts were compared based on the concentration changes of 4-NP as a function of time (Figure 5c). The reaction would not happen in the absence of catalysts. It was observed that the decrease in the concentration of 4-NP over the catalysts followed the order: NiPt-0.6% > NiPt-0.2% > NiPt-1.0% > Ni. The concentration of 4-NP decreased to 10% of the initial value within 120 seconds over NiPt-0.6% catalysts. It was noted that the optimal catalysts were NiPt-0.6% rather than NiPt-1.0% with a higher Pt content. This suggested that the activity of Ni catalysts could be enhanced by incorporating an appropriate amount of Pt. The excessive amount of Pt had a negative effect on the catalytic activity in the reduction of 4-NP to 4-AP. These findings were in agreement with the catalytic behaviors of AuCu alloy reported in a recent literature [39]. In that work, the reaction pathway for reduction of nitroaromatics to anilines was adjusted by tailoring the compositions of bimetallic nanocatalysts. A similar volcano curve associated with the yield of aniline was observed with the change of Cu contents.

The volcano curve behavior indicated that the chemical compositions of the bimetallic nanocatalysts were important parameters for affecting the catalytic activity.

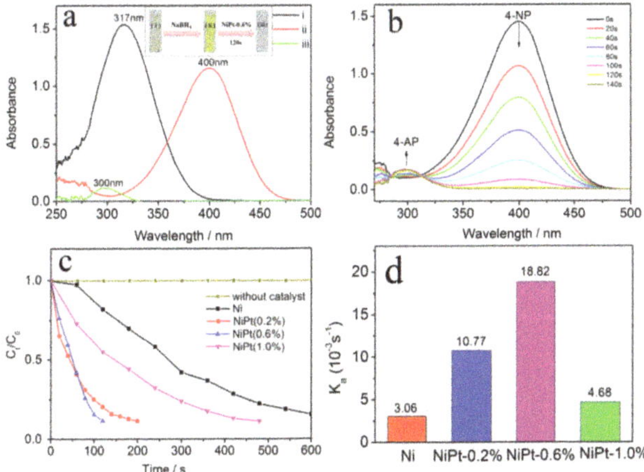

Figure 5. (**a**) Ultraviolet-visible (UV-VIS) absorption spectra of the solution before (i) and after (ii) the addition of $NaBH_4$ and (iii) after the addition of NiPt-0.6% catalyst; (**b**) time-dependent UV-VIS absorption spectra of the reduction of 4-NP over the NiPt-0.6% catalyst in aqueous solution at room temperature; (**c**) reduction of 4-NP over different catalysts as a function of time; (**d**) apparent rate constants (K_a) of the reactions in the presence of supported catalysts.

The reduction rate of 4-NP was independent of the concentration of reducing agent ($NaBH_4$) since the initial concentration of $NaBH_4$ greatly exceeded that of 4-NP (~100 times). Thus, the reduction process could be described with pseudo-first-order kinetics with respect to the concentration of 4-NP [40]:

$$-\ln\left(C_t/C_0\right) = K_a t \tag{3}$$

where C_0 and C_t (unit in mM) were the concentrations of 4-NP at the beginning and at a certain time, respectively. Figure 5d showed the apparent rate constants (K_a) for different catalysts, which were calculated according to equation 3. The values of K_a for NiPt-0.2%, NiPt-0.6%, and NiPt-1.0% catalysts were 0.01077, 0.01882, and 0.00468 s^{-1}, respectively, all of which were higher than that of the Ni catalysts (0.00306 s^{-1}). This could be caused by the strong interactions between Ni and Pt. The electronic structures of Ni were modified by incorporation of Pt, which was consistent with the recent report on Ir-promoted Ni catalysts for hydrogenation reaction [18]. As to the different activities of NiPt catalysts, it might be due to the different exposed crystal planes of Ni NPs and Pt NPs. As shown in Figure 2c, the NiPt-0.6% catalysts showed the Ni (200) plane besides the Ni (111) one. The crystal plane-dependent activity has been confirmed in metal-supported oxide catalysts [41]. Further studies on the correlations were required by constructing a less complicated model catalyst.

In addition, the K_a values of various catalysts were compared in Table 3. It was seen that the NiPt-0.6% catalyst showed much larger K_a value than the reported nickel catalysts [4,6,14]. Importantly, this catalyst exhibited higher activity than noble-metal and bimetallic ones such as DPNs (dendritic Pt nanoparticles) [11], reduced graphene oxide (RGO)/PtNi [35], Pt/γ-Al_2O_3 [40] and Ni-Pt [42]. The comparisons further verified the superior catalytic activity of the optimal NiPt catalysts for 4-NP reduction and highlighted the very limited use of noble-metal Pt in the as-prepared high-performance catalysts.

Table 3. Comparisons of the apparent rate constant (K_a) for 4-NP reduction over various catalysts.

Catalysts	Reaction Conditions [1]	K_a [2]/10^{-3} s^{-1}	References
NiPt-0.6% (Ni: 36.98 wt % Pt: 0.347 wt %)	15.5 mg, 2 mM, 25 °C, 0.25 M	18.82	This work
$Pd_{0.05}$/G	4 mg, 0.3 mM, 25 °C, 0.1 M	36.5	[1]
Ir/IrO_x	-, 20 mM, 25 °C, 0.2 M	2.57	[2]
230 nm Ni/SiO_2 MHMs (Ni: 14.6 wt %)	3 mg, 5 mM, 25 °C, 0.2 M	4.5	[4]
Ni NPs	3 mg, 0.1mM, 20 °C, 0.2 M	2.7	[6]
Pd-Fe_3O_4 (1.2 wt %)	10 mg, 21.56 mM, 25 °C, 0.1 MPa	-	[7]
DPNs	-, 2 mM, 25 °C, 0.3 M	0.75	[11]
Ni/TiO_2	400 mg, -, 100 °C, 1.5 MPa	-	[14]
Ni/SiO_2@Au MHMs	4 mg, 5 mM, 25 °C, 0.2 M	10	[15]
RGO/PtNi (25:75)	3 mg, 5 mM, 25 °C, 1.5 M	1.12	[35]
TAC-Ag-1.0	0.004 mg, 0.103 mM, 25 °C, 0.3 M	5.19	[38]
Au-Cu alloy NP	50 mg, -, 60 °C, -	-	[39]
Pt/γ-Al_2O_3 (2.7 wt %)	0.5 mg, 1 mM, 22 °C, 0.1M	0.53	[40]
Ni-Pt (96:4)	0.004 mg, 0.085 mM, 25 °C, 0.012 M	1.93	[42]
AuNPs/SNTs	8 mg, 0.12 mM, 25 °C, 0.005 M	10.64	[43]
AuNP/CeO_2 (Au: 0.031 mg)	10 mg, 0.12 mM, 25 °C, 0.005 M	2.25	[44]

[1] Reaction conditions follow the order of amount of catalyst, the initial concentration of 4-NP, temperature, H_2 pressure/the concentration of $NaBH_4$; [2] K_a: Apparent rate constant.

Generally, the formation of a bimetallic structure can change the electronic states of metals, particularly the d-band center of the involved metal atoms, which is highly relevant to their catalytic activity [45]. The variations of the d-band in metals have been theoretically studied using density functional theory (DFT) computations assuming some specific extended-surface models [46,47]. Recently, the theory has also been applied to bimetallic systems to correlate the changes in the electronic states with the catalytic activities in a variety of reactions [48]. With regard to the NiPt NPs studied in this work, Ni was reduced *in situ* and enriched on the surface from a hybrid of LDH/carbon precursor. The pristine Ni NPs showed the activity towards the reduction of 4-NP. Upon introducing Pt to the Ni catalysts, the local electronic structures of the Ni were affected, and the electron density on the Ni was changed, which was dependent on the bimetallic compositions [42]. It was found that the NiPt-0.6% catalysts had higher activity than the NiPt-1.0% with a larger content of Pt. It could be due to the different exposed crystal planes of Ni NPs (Figure 2c).

On the other hand, small nanoparticles and excellent dispersions resulted in high activity in many cases [49,50]. The sizes of nickel crystallites as estimated by TEM decreased with increasing amount of Pt. No obvious agglomeration of Ni NPs was observed although the Ni loading was very high (36%–39%). This highlighted the advantages of the precursor method from LDH/carbon hybrid owing to the lattice anchoring effect of metal ions in the layers of LDH [22,25]. The Pt-modified Ni NPs were more active for adsorbing and cleaving H_2 molecules than the pristine Ni NPs during reduction of 4-NP, owing to a higher electron density of NiPt active sites [51]. The produced active hydrogen atoms reacted with 4-NP on the active sites to generate 4-AP. A reaction pathway was shown in Figure 6. First, $NaBH_4$ rapidly reacted with water to release H_2, in which the sodium metaborate ($NaBO_2$) was formed as a byproduct. Then, the H_2 split H–H bond into H atoms, which adsorbed on the surface of metal nanoparticles. There was no doubt that the increase in chemisorbed H atoms on NiPt nanoparticles could improve the reducibility. It was accepted that the more hydrogen adsorbed, the higher the catalytic activity exhibited [18]. The negatively-charged hydrogen in the metal–hydrogen bond attacked the positively charged nitrogen within the nitro group of 4-NP. Subsequently the nitro group was reduced to the nitroso group intermediate, followed by the reaction of two hydrogen atoms to form hydroxylamine. Finally, the hydroxylamine was further reduced to amino group to obtain 4-AP.

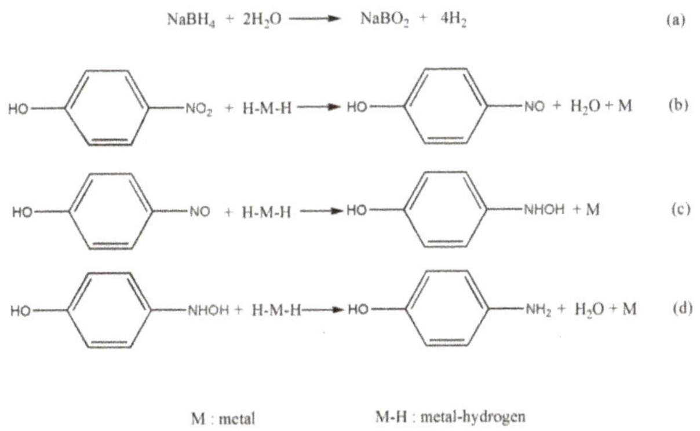

$$NaBH_4 + 2H_2O \longrightarrow NaBO_2 + 4H_2 \quad \text{(a)}$$

$$HO{-}C_6H_4{-}NO_2 + \text{H-M-H} \longrightarrow HO{-}C_6H_4{-}NO + H_2O + M \quad \text{(b)}$$

$$HO{-}C_6H_4{-}NO + \text{H-M-H} \longrightarrow HO{-}C_6H_4{-}NHOH + M \quad \text{(c)}$$

$$HO{-}C_6H_4{-}NHOH + \text{H-M-H} \longrightarrow HO{-}C_6H_4{-}NH_2 + H_2O + M \quad \text{(d)}$$

M : metal M-H : metal-hydrogen

Figure 6. Reaction pathways for the reduction of 4-NP to 4-AP: (**a**) $NaBH_4$ hydrolyzed to release H_2; (**b**) H_2 molecules split to H atoms on the surface of metal NPs and reacted with 4-NP; (**c**) H atoms reacted with the nitrosophenol intermediate to form hydroxylamine; (**d**) hydroxylamine was further reduced to the final product 4-AP. H–M–H represents the split of H_2 molecules on the surface of metal NPs.

The native magnetic property of the nanocatalysts caused by Ni made them economically and easily separated after reactions using an external magnet (Figure 7), and the catalysts were conveniently recycled and reused in a fresh solution.

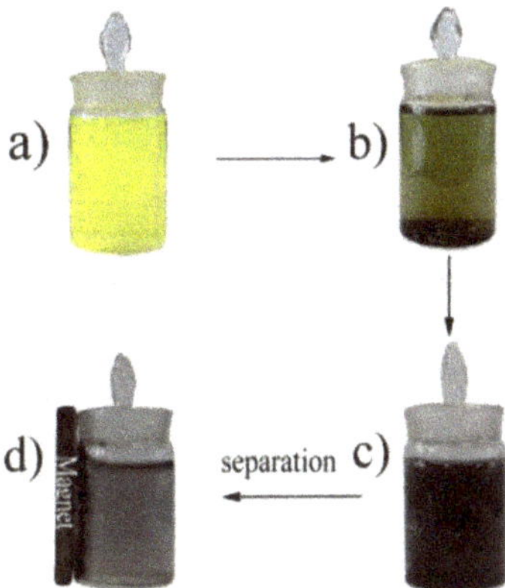

Figure 7. Separation of NiPt-0.6% from solution by a magnet: (**a**) before addition of the catalyst; (**b**) after addition of the catalyst; (**c**) the suspension solution after reaction; (**d**) separation of solid catalysts with a magnet.

3. Experimental Section

3.1. Materials

Hexachloroplatinic acid ($H_2PtCl_6 \cdot 6H_2O$, 99.9%) was purchased from Sinopharm Chemical Reagent Co., Ltd. (Shanghai, China). Glucose anhydrous and $NaBH_4$ were purchased from Kermel Chemical Reagent Co., Ltd. (Tianjin, China). 4-nitrophenol, received from Macklin Biochemical Co., Ltd. (Shanghai, China), was used without further purification. Other chemicals and reagents were all analytical reagent grade and used as received without further purification. Deionized water was used throughout the process.

3.2. *Synthesis of NiAl-LDH Precursor*

A mixture of $Ni(NO_3)_2 \cdot 6H_2O$ and $Al(NO_3)_3 \cdot 9H_2O$ were dissolved in 30 mL of deionized water to form a clear salt solution ($[Ni^{2+}]$ = 0.2 M, $[Al^{3+}]$ = 0.1 M). The salt solution was rapidly poured into a 30 mL NaOH and Na_2CO_3 solution ($[OH^-] = 1.6[Ni^{2+} + Al^{3+}]$, $[CO_3{}^{2-}] = 2[Al^{3+}]$) within several tens of seconds under vigorous stirring. The mixture solution was further stirred for 10 min at room temperature. The resulting suspension was then centrifuged and re-dispersed in deionized water for five cycles to obtain LDH precipitate. The resulting precipitate was transferred and dispersed into a Teflon-lined autoclave (100 mL total volume), and 80 mL of deionized water was added to it. The autoclave was then tightly sealed and maintained at 150 °C for 10 h. The resulting bluish green suspension was directly vacuum-freeze-dried overnight to collect the solid NiAl-LDH powder.

3.3. *Synthesis of LDH/Carbon Composites and Supported NiPt Catalysts*

A certain amount of $H_2PtCl_6 \cdot 6H_2O$ was added to 30 mL of glucose ($C_6H_{12}O_6$) solution ($[C_6H_{12}O_6] = 2.5[Ni^{2+} + Al^{3+}]$). The precipitate obtained via the above-mentioned room-temperature co-precipitation was dispersed in the mixture solution. The resultant solution was transferred into an autoclave with the same fill volume and maintained at the same temperature (150 °C) for the same period of time (10 h). Subsequently, the product was centrifuged and washed with deionized water and ethanol five times. Finally, the precipitate was freeze-dried overnight to obtain LDH/C composites and Pt/LDH-C. The resulting composites were loaded into an alumina boat and calcined in a tube furnace in a flowing N_2 atmosphere. The furnace was heated to a preset temperature at a rate of 5 °C/min, and the temperature was maintained for 2 h. After the reaction, the furnace was naturally cooled to room temperature. The obtained black product was denoted as NiPt-x, where x is the theoretical loading percentage of Pt (wt %).

3.4. *Characterizations*

UV-VIS absorption spectra were recorded using a (TU-1810 spectrometer, Purkinje, Inc., Beijing, China. TEM and high-resolution TEM (HRTEM) images were obtained using a JEM-2100 microscope (JEOL, Tokyo, Japan) with an Oxford INCA detector (Oxford-instruments, Shanghai, China) operating at 200 kV. The compositions of the as-prepared catalysts were determined through inductively coupled plasma-atomic emission spectroscopy (ICP-AES) (ICPS-7500 Spectrometer, Shimadzu, Inc., Beijing, China) Powder XRD analyses were performed using a Bruker D8 Advanced diffractometer (Bruker, Beijing, China) with Cu Kα radiation, and the scanning angle 2θ ranged from 5° to 80°. XPS measurements were performed using a VG ESCALAB250 (Thermo Fisher Scientific, MA, USA) with Al K_α radiation ($h\nu$ = 1486.6 eV). The specific surface areas were measured via N_2-adsorption at 77 K using a static volumetric Quantachrome Autosorb-1C-VP Analyser (Quantachrome, Shanghai, China). The specific surface area was calculated using a Brunauer–Emmett–Teller (BET) method.

3.5. *Catalytic Evaluation*

Catalytic reduction reactions of 4-nitrophenols were conducted at room temperature (25 °C) in the presence of NiPt catalysts with varying Pt content. Typically, aqueous solutions of 4-NP (2 mM) and $NaBH_4$ (0.25 M) were freshly prepared. NiPt catalyst in the amount of 15.5 mg was dispersed into 40 mL of 4-NP and $NaBH_4$ mixture solutions under continuous stirring. To evaluate the reaction progress, approximately 1 mL of solution was taken out of the reaction mixture at specified time intervals and subsequently diluted 20 times with deionized water. This step was followed by recording of the UV-VIS spectra of the solution to examine the concentration of 4-NP by monitoring the adsorption peak at 400 nm.

4. Conclusions

The supported NiPt nanocatalysts were synthesized by an *in situ* reduction strategy from Pt@LDH/carbon composites precursor. The precursor method ensured the formation of nanosized (~10 nm) and aggregate-free Ni NPs. The incorporation of Pt to the pristine Ni catalysts led to the enhanced activity towards the reduction of 4-NP to 4-AP because the modification of Pt affected the electronic structures of Ni. The optimal NiPt catalysts with an appropriate amount of Pt showed excellent activity in the reduction of 4-NP. The apparent rate constant (K_a) reached 18.82×10^{-3} s^{-1}, which was 6.2 times higher than that of Ni catalysts and much higher than most of reported catalysts in the literaures. The present method for bimetallic catalysts could be extended to other compositions owing to the adjustable components in LDHs to improve the catalytic activity of the monometallic ones and simultaneously to minimize the utilization of noble metals.

Acknowledgments: This work was supported by the 973 Program (Grant 2014CB932104), the National Natural Science Foundation of China (NSFC), the Beijing Natural Science Foundation (Grant 2152022), the Fundamental Research Funds for the Central Universities (Grant YS1406) and Jiangsu Key Laboratory of Advanced Catalytic Materials and Technology (Grant BM2012110).

Author Contributions: Xu Xiang and Bing Zhang conceived the idea and designed the experiments. Huishan Shang, Kecheng Pan and Lu Zhang were involved in the synthesis, characterization and catalytic evaluation of catalysts. Xu Xiang, Huishan Shang and Bing Zhang conducted the discussion, interpreted the data and wrote the manuscript. All authors read the manuscript and agreed the submission.

Conflicts of Interest: The authors declare no conflict of interest.

References

1. Sun, J.; Fu, Y.; He, G.; Sun, X.; Wang, X. Catalytic hydrogenation of nitrophenols and nitrotoluenes over a palladium/graphene nanocomposite. *Catal. Sci. Technol.* **2014**, *4*, 1742–1748. [CrossRef]
2. Xu, D.; Diao, P.; Jin, T.; Wu, Q.; Liu, X.; Guo, X.; Gong, H.; Li, F.; Xiang, M.; Ronghai, Y. Iridium oxide nanoparticles and iridium/iridium oxide nanocomposites: Photochemical fabrication and application in catalytic reduction of 4-nitrophenol. *ACS Appl. Mater. Interfaces* **2015**, *7*, 16738–16749. [CrossRef] [PubMed]
3. Liu, W.-J.; Tian, K.; Jiang, H. One-pot synthesis of Ni–NiFe$_2$O$_4$/carbon nanofiber composites from biomass for selective hydrogenation of aromatic nitro compounds. *Green Chem.* **2015**, *17*, 821–826. [CrossRef]
4. Niu, Z.; Zhang, S.; Sun, Y.; Gai, S.; He, F.; Dai, Y.; Li, L.; Yang, P. Controllable synthesis of Ni/SiO$_2$ hollow spheres and their excellent catalytic performance in 4-nitrophenol reduction. *Dalton Trans.* **2014**, *43*, 16911–16918. [CrossRef] [PubMed]
5. Mohan, V.; Pramod, C.; Suresh, M.; Reddy, K.H.P.; Raju, B.D.; Rao, K.R. Advantage of Ni/SBA-15 catalyst over Ni/MgO catalyst in terms of catalyst stability due to release of water during nitrobenzene hydrogenation to aniline. *Catal. Commun.* **2012**, *18*, 89–92. [CrossRef]
6. Jiang, Z.; Xie, J.; Jiang, D.; Wei, X.; Chen, M. Modifiers-assisted formation of nickel nanoparticles and their catalytic application to *p*-nitrophenol reduction. *CrystEngComm* **2013**, *15*, 560–569. [CrossRef]
7. Zhang, D.; Chen, L.; Ge, G. A green approach for efficient *p*-nitrophenol hydrogenation catalyzed by a Pd-based nanocatalyst. *Catal. Commun.* **2015**, *66*, 95–99. [CrossRef]
8. Kumar, A.; Kumar, P.; Joshi, C.; Manchanda, M.; Boukherroub, R.; Jain, S.L. Nickel decorated on phosphorous-doped carbon nitride as an efficient photocatalyst for reduction of nitrobenzenes. *Nanomaterials* **2016**, *6*. [CrossRef]
9. Deka, P.; Deka, R.C.; Bharali, P. *In situ* generated copper nanoparticle catalyzed reduction of 4-nitrophenol. *New J. Chem.* **2014**, *38*, 1789–1793. [CrossRef]
10. Gao, Y.; Ding, X.; Zheng, Z.; Cheng, X.; Peng, Y. Template-free method to prepare polymer nanocapsules embedded with noble metal nanoparticles. *Chem. Commun.* **2007**, *36*, 3720–3722. [CrossRef] [PubMed]
11. Wang, J.; Zhang, X.-B.; Wang, Z.-L.; Wang, L.-M.; Xing, W.; Liu, X. One-step and rapid synthesis of "clean" and monodisperse dendritic Pt nanoparticles and their high performance toward methanol oxidation and *p*-nitrophenol reduction. *Nanoscale* **2012**, *4*, 1549–1552. [CrossRef] [PubMed]

12. Baruah, B.; Gabriel, G.J.; Akbashev, M.J.; Booher, M.E. Facile synthesis of silver nanoparticles stabilized by cationic polynorbornenes and their catalytic activity in 4-nitrophenol reduction. *Langmuir* **2013**, *29*, 4225–4234. [CrossRef] [PubMed]
13. Gangula, A.; Podila, R.; Karanam, L.; Janardhana, C.; Rao, A.M. Catalytic reduction of 4-nitrophenol using biogenic gold and silver nanoparticles derived from *Breynia rhamnoides*. *Langmuir* **2011**, *27*, 15268–15274. [CrossRef] [PubMed]
14. Wu, Z.; Chen, J.; Di, Q.; Zhang, M. Size-controlled synthesis of a supported Ni nanoparticle catalyst for selective hydrogenation of *p*-nitrophenol to *p*-aminophenol. *Catal. Commun.* **2012**, *18*, 55–59. [CrossRef]
15. Zhang, S.; Gai, S.; He, F.; Dai, Y.; Gao, P.; Li, L.; Chen, Y.; Yang, P. Uniform Ni/SiO_2@Au magnetic hollow microspheres: Rational design and excellent catalytic performance in 4-nitrophenol reduction. *Nanoscale* **2014**, *6*, 7025–7032. [CrossRef] [PubMed]
16. Watanabe, M.; Yamashita, H.; Chen, X.; Yamanaka, J.; Kotobuki, M.; Suzuki, H.; Uchida, H. Nano-sized Ni particles on hollow alumina ball: Catalysts for hydrogen production. *Appl. Catal. B Environ.* **2007**, *71*, 237–245. [CrossRef]
17. Wang, D.; Astruc, D. The golden age of transfer hydrogenation. *Chem. Rev.* **2015**, *115*, 6621–6686. [CrossRef] [PubMed]
18. Lin, W.; Cheng, H.; He, L.; Yu, Y.; Zhao, F. High performance of Ir-promoted Ni/TiO_2 catalyst toward the selective hydrogenation of cinnamaldehyde. *J. Catal.* **2013**, *303*, 110–116. [CrossRef]
19. Wang, S.; Lin, W.; Zhu, Y.; Xie, Y.; McCormick, J.R.; Huang, W.; Chen, J.G. Pd-based bimetallic catalysts prepared by replacement reactions. *Catal. Lett.* **2007**, *114*, 169–173. [CrossRef]
20. Zhao, Z.-F.; Wu, Z.-J.; Zhou, L.-X.; Zhang, M.-H.; Li, W.; Tao, K.-Y. Synthesis of a nano-nickel catalyst modified by ruthenium for hydrogenation and hydrodechlorination. *Catal. Commun.* **2008**, *9*, 2191–2194. [CrossRef]
21. Nishikawa, J.; Miyazawa, T.; Nakamura, K.; Asadullah, M.; Kunimori, K.; Tomishige, K. Promoting effect of Pt addition to Ni/CeO_2/Al_2O_3 catalyst for steam gasification of biomass. *Catal. Commun.* **2008**, *9*, 195–201. [CrossRef]
22. Xiang, X.; Bai, L.; Li, F. Formation and catalytic performance of supported Ni nanoparticles via self-reduction of hybrid NiAl-LDH/C composites. *AIChE J.* **2010**, *56*, 2934–2945. [CrossRef]
23. Wang, Y.; He, W.; Wang, L.; Yang, J.; Xiang, X.; Zhang, B.; Li, F. Highly active supported Pt nanocatalysts synthesized by alcohol reduction towards hydrogenation of cinnamaldehyde: Synergy of metal valence and hydroxyl groups. *Chem. Asian J.* **2015**, *10*, 1561–1570. [CrossRef] [PubMed]
24. Xiang, X.; He, W.; Xie, L.; Li, F. A mild solution chemistry method to synthesize hydrotalcite-supported platinum nanocrystals for selective hydrogenation of cinnamaldehyde in neat water. *Catal. Sci. Technol.* **2013**, *3*, 2819–2827. [CrossRef]
25. Xiang, X.; Hima, H.I.; Wang, H.; Li, F. Facile Synthesis and Catalytic Properties of nickel-based mixed-metal oxides with mesopore networks from a novel hybrid composite precursor. *Chem. Mater.* **2007**, *20*, 1173–1182. [CrossRef]
26. Xiang, X.; Li, F.; Huang, Z. Recent advances in layered double hydroxide-based materials as versatile photocatalysts. *Rev. Adv. Sci. Eng.* **2014**, *3*, 158–171. [CrossRef]
27. Gawande, M.B.; Rathi, A.K.; Branco, P.S.; Nogueira, I.D.; Velhinho, A.; Shrikhande, J.J.; Indulkar, U.U.; Jayaram, R.V.; Ghumman, C.A.A.; Bundaleski, N. Regio-and chemoselective reduction of nitroarenes and carbonyl compounds over recyclable magnetic ferrite-nickel nanoparticles (Fe_3O_4-Ni) by using glycerol as a hydrogen source. *Chem. Eur. J.* **2012**, *18*, 12628–12632. [CrossRef] [PubMed]
28. Wang, Q.; Wang, Y.; Zhao, Y.; Zhang, B.; Yunyin, N.; Xiang, X.; Chen, R. Fabricating roughened surfaces on halloysite nanotubes via alkali etching for deposition of high-efficiency Pt nanocatalysts. *CrystEngComm* **2015**, *17*, 3110–3116. [CrossRef]
29. Gawande, M.B.; Branco, P.S.; Varma, R.S. Nano-magnetite (Fe_3O_4) as a support for recyclable catalysts in the development of sustainable methodologies. *Chem. Soc. Rev.* **2013**, *42*, 3371–3393. [CrossRef] [PubMed]
30. Stassi, J.P.; Zgolicz, P.D.; de Miguel, S.R.; Scelza, O.A. Formation of different promoted metallic phases in PtFe and PtSn catalysts supported on carbonaceous materials used for selective hydrogenation. *J. Catal.* **2013**, *306*, 11–29. [CrossRef]

31. Bera, P.; Priolkar, K.; Gayen, A.; Sarode, P.; Hegde, M.; Emura, S.; Kumashiro, R.; Jayaram, V.; Subbanna, G. Ionic dispersion of Pt over CeO_2 by the combustion method: Structural investigation by XRD, TEM, XPS, and EXAFS. *Chem. Mater.* **2003**, *15*, 2049–2060. [CrossRef]
32. Zhou, X.-W.; Zhang, R.-H.; Zhou, Z.-Y.; Sun, S.-G. Preparation of PtNi hollow nanospheres for the electrocatalytic oxidation of methanol. *J. Power Sources* **2011**, *196*, 5844–5848. [CrossRef]
33. Pawelec, B.; Damyanova, S.; Arishtirova, K.; Fierro, J.; Petrov, L. Structural and surface features of PtNi catalysts for reforming of methane with CO_2. *Appl. Catal. A* **2007**, *323*, 188–201. [CrossRef]
34. Haber, J.A.; Cai, Y.; Jung, S.; Xiang, C.; Mitrovic, S.; Jin, J.; Bell, A.T.; Gregoire, J.M. Discovering Ce-rich oxygen evolution catalysts, from high throughput screening to water electrolysis. *Energy Environ. Sci.* **2014**, *7*, 682–688. [CrossRef]
35. Sahoo, P.K.; Panigrahy, B.; Bahadur, D. Facile synthesis of reduced graphene oxide/Pt–Ni nanocatalysts: Their magnetic and catalytic properties. *RSC Adv.* **2014**, *4*, 48563–48571. [CrossRef]
36. Karelovic, A.; Ruiz, P. Improving the hydrogenation function of Pd/γ-Al_2O_3 catalyst by Rh/γ-Al_2O_3 addition in CO_2 methanation at low temperature. *ACS Catal.* **2013**, *3*, 2799–2812. [CrossRef]
37. Dhar, J.; Patil, S. Self-assembly and catalytic activity of metal nanoparticles immobilized in polymer membrane prepared via layer-by-layer approach. *ACS Appl. Mater. Interfaces* **2012**, *4*, 1803–1812. [CrossRef] [PubMed]
38. Rashid, M.H.; Mandal, T.K. Synthesis and catalytic application of nanostructured silver dendrites. *J. Phys. Chem. C* **2007**, *111*, 16750–16760. [CrossRef]
39. Xiao, Q.; Sarina, S.; Waclawik, E.R.; Jia, J.; Chang, J.; Riches, J.D.; Wu, H.; Zheng, Z.; Zhu, H. Alloying gold with copper makes for a highly selective visible-light photocatalyst for the reduction of nitroaromatics to anilines. *ACS Catal.* **2016**, *6*, 1744–1753. [CrossRef]
40. Dandapat, A.; Jana, D.; De, G. Synthesis of thick mesoporous γ-alumina films, loading of Pt nanoparticles, and use of the composite film as a reusable catalyst. *ACS Appl. Mater. Interfaces* **2009**, *1*, 833–840. [CrossRef] [PubMed]
41. Na, T.; Liu, J.; Wenjie, S. Tuning the shape of ceria nanomaterials for catalytic applications. *Chin. J. Catal.* **2013**, *34*, 838–850.
42. Ghosh, S.K.; Mandal, M.; Kundu, S.; Nath, S.; Pal, T. Bimetallic Pt–Ni nanoparticles can catalyze reduction of aromatic nitro compounds by sodium borohydride in aqueous solution. *Appl. Catal. A* **2004**, *268*, 61–66. [CrossRef]
43. Zhang, Z.; Shao, C.; Zou, P.; Zhang, P.; Zhang, M.; Mu, J.; Guo, Z.; Li, X.; Wang, C.; Liu, Y. *In situ* assembly of well-dispersed gold nanoparticles on electrospun silica nanotubes for catalytic reduction of 4-nitrophenol. *Chem. Commun.* **2011**, *47*, 3906–3908. [CrossRef] [PubMed]
44. Zhang, J.; Chen, G.; Chaker, M.; Rosei, F.; Ma, D. Gold nanoparticle decorated ceria nanotubes with significantly high catalytic activity for the reduction of nitrophenol and mechanism study. *Appl. Catal. B* **2013**, *132*, 107–115. [CrossRef]
45. Stamenkovic, V.R.; Mun, B.S.; Arenz, M.; Mayrhofer, K.J.; Lucas, C.A.; Wang, G.; Ross, P.N.; Markovic, N.M. Trends in electrocatalysis on extended and nanoscale Pt-bimetallic alloy surfaces. *Nat. Mater.* **2007**, *6*, 241–247. [CrossRef] [PubMed]
46. Nørskov, J.K.; Bligaard, T.; Logadottir, A.; Bahn, S.; Hansen, L.B.; Bollinger, M.; Bengaard, H.; Hammer, B.; Sljivancanin, Z.; Mavrikakis, M. Universality in heterogeneous catalysis. *J. Catal.* **2002**, *209*, 275–278. [CrossRef]
47. Greeley, J.; Nørskov, J.K.; Mavrikakis, M. Electronic structure and catalysis on metal surfaces. *Annu. Rev. Phys. Chem.* **2002**, *53*, 319–348. [CrossRef] [PubMed]
48. Zhou, W.P.; Lewera, A.; Larsen, R.; Masel, R.I.; Bagus, P.S.; Wieckowski, A. Size effects in electronic and catalytic properties of unsupported palladium nanoparticles in electrooxidation of formic acid. *J. Phys. Chem. B* **2006**, *110*, 13393–13398. [CrossRef] [PubMed]
49. Nie, Y.; Chen, S.; Ding, W.; Xie, X.; Zhang, Y.; Wei, Z. Pt/C trapped in activated graphitic carbon layers as a highly durable electrocatalyst for the oxygen reduction reaction. *Chem. Commun.* **2014**, *50*, 15431–15434. [CrossRef] [PubMed]

50. Yang, X.; Chen, D.; Liao, S.; Song, H.; Li, Y.; Fu, Z.; Su, Y. High-performance Pd–Au bimetallic catalyst with mesoporous silica nanoparticles as support and its catalysis of cinnamaldehyde hydrogenation. *J. Catal.* **2012**, *291*, 36–43. [CrossRef]
51. Joseph, T.; Kumar, K.V.; Ramaswamy, A.; Halligudi, S. Au–Pt nanoparticles in amine functionalized MCM-41: Catalytic evaluation in hydrogenation reactions. *Catal. Commun.* **2007**, *8*, 629–634. [CrossRef]

 *nanomaterials*

Article

Hydrothermal Synthesis of Ultrasmall Pt Nanoparticles as Highly Active Electrocatalysts for Methanol Oxidation

Wenhai Ji [1], Weihong Qi [1,2,3,*], Shasha Tang [1], Hongcheng Peng [1] and Siqi Li [1]

1. School of Materials Science and Engineering, Central South University, Changsha 410083, China; jwh1988@csu.edu.cn (W.J.); tang0360@ntu.edu.sg (S.T.); 133111028@csu.edu.cn (H.P.); 9901090160@csu.edu.cn (S.L.)
2. Institute for Materials Microstructure, Central South University, Changsha 410083, China
3. Key Laboratory of Non-Ferrous Materials Science and Engineering, Ministry of Education, Changsha 410083, China

* Author to whom correspondence should be addressed; qiwh216@csu.edu.cn; Tel.: +86-731-88876692.

Academic Editors: Hermenegildo García and Sergio Navalón
Received: 31 October 2015; Accepted: 25 November 2015; Published: 8 December 2015

Abstract: Ultrasmall nanoparticles, with sizes in the 1–3 nm range, exhibit unique properties distinct from those of free molecules and larger-sized nanoparticles. Demonstrating that the hydrothermal method can serve as a facile method for the synthesis of platinum nanoparticles, we successfully synthesized ultrasmall Pt nanoparticles with an average size of 2.45 nm, with the aid of poly(vinyl pyrrolidone) (PVP) as reducing agents and capping agents. Because of the size effect, these ultrasmall Pt nanoparticles exhibit a high activity toward the methanol oxidation reaction.

Keywords: hydrothermal method; ultrasmall nanoparticles; PVP; methanol oxidation

---

1. Introduction

Platinum has been widely used in many applications, especially for catalysis and fuel cell technology, due to its ability to facilitate both oxidation and reduction reactions [1–5]. Size and shape play an important role in determining properties and potential applications of nanomaterials. Thus, downsizing these nanomaterials can provide a great opportunity to achieve a high surface-to-volume ratio and thus enhance the metal utilization in noble metal–based catalysts or electrocatalysts. Ultrasmall nanoparticles (USNPs), with sizes in the 1–3 nm range, exhibit unique properties distinct from those of free molecules and larger-sized nanoparticles.

Pt USNPs have also been reported. By using alcohols as reductants in the presence of poly(vinyl pyrrolidone) (PVP) in a refluxing aqueous system, Pt USNPs were obtained [6]. The average size of Pt USNPs could be controlled from 1.9 to 3.3 nm by changing the alcohol or the concentrations of the reagents. Photochemical methodology also could be used as a very effective way to produce Pt USNPs [7]. It involved visible light as a reaction trigger and platinum acetylacetonate and thioglycolic acid as the only chemical reactants. The average diameter of platinum nanoparticles was 1.0 ± 0.3 nm. Li *et al.* reported the synthesis of monodisperse Pt USNPs stabilized with peptides in aqueous solution at room temperature [8]. The specifically selected peptide molecule, P7A, is able to bind to the surface of the Pt NPs and regulate the nucleation and growth rates, affording monodisperse Pt NPs with sizes in the 1.7–3.5 nm range. Pt USNPs have also been synthesized with hydrophobic ligands. Uniform 2-nm-sized Pt USNPs were obtained by the decomposition of platinum dibenzylideneacetone under mild conditions in the presence of n-octylsilane [9]. Lim *et al.* successfully synthesized Pt USNPs with an average size of 1.9 nm through the reaction of a Pt salt precursor with water, without the aid of any exotic reducing agents and organic capping molecules [10]. The *in situ* synthesis of Pt USNPs at room temperature using poly(4-vinyl phenol) (PVPh) as both the reducing as well as the stabilizing agent in aqueous alkaline solution has been reported [11]. Transmission electron microscopic analysis confirms the formation of ultra-small spherical Pt USNPs from 1.6 to 2.2 nm in diameter with a high degree of monodispersity depending on the ratio of PVPh to platinum salt concentrations used in a single reaction.

Specific organic capping agents or stabilizers have been exploited in order to restrict nanocrystal growth as well as to provide a barrier to agglomeration. Notably, PVP has received special attention due to its high chemical stability, nontoxicity, and excellent solubility in many polar solvents. Hydroxyl end groups of PVP can be applied as a general strategy for the kinetically controlled synthesis of nanoplates made of noble metals such as Ag, Pd, Au, and Pt [12,13]. PVP (see Scheme 1) has an affinity for long-chain alcohols owing to its hydroxyl end groups; it can serve as a reducing agent. As a major advantage over alcohols with short alkyl chains commonly used in chemical synthesis, the PVP can be used as both reductant and stabilizer.

The hydrothermal method has been used in the synthesis of semiconductor nanocrystals and nanocomposites for a long time, but its role remains largely unexplored in the synthesis of noble-metal colloids. Here we demonstrate that Pt USNPs can be synthesized through a hydrothermal method with PVP. To the best of our knowledge, the present work may be the first report on synthesizing Pt USNPs in aqueous solutions by means of a simple hydrothermal method.

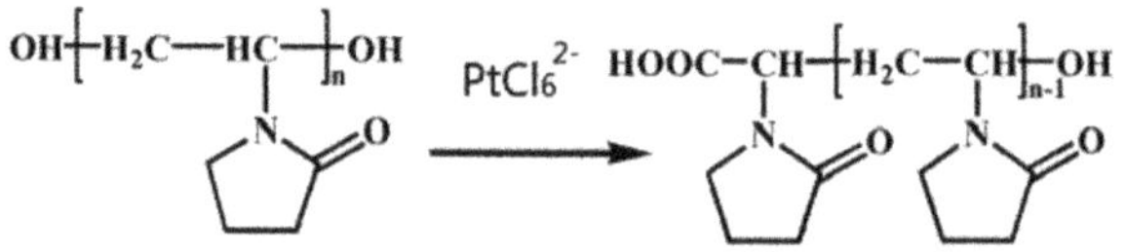

Scheme 1. Formulas of OH-terminated poly(vinyl pyrrolidone) (PVP) and the structural variation in the effect of $PtCl_6^{2-}$ ions.

2. Results and Discussion

We synthesized Pt USNPs by reacting K_2PtCl_6 with PVP in a closed autoclave in an oven at 160 °C for 8 h. Transmission electron microscope (TEM) images in Figure 1a,b show that the Pt USNPs have an average size around 2.45 nm. The high resolution TEM image of the crystal structure of the Pt USNPs in Figure 1c displays a well-defined crystal lattice. The inset image in Figure 1c shows that a d spacing between adjacent lattice planes of 0.224 nm corresponds to the {111} planes, and 0.195 nm corresponds to the {200} planes. TEM studies and a size histogram of Pt USNPs obtained by counting 200 particles (Figure 1d) show that these Pt USNPs typically have nearly spherical shapes, with a narrow size distribution. The formation of Pt USNPs synthesized in the hydrothermal method is due to the presence of PVP. The OH-terminated PVP can be applied as the reductant for the synthesis of nanoparticles made of other noble metals (such as Ag, Pd, Au, and Pt) [12–15]. Compared with alcohols (commonly used in the chemical synthesis), the PVP can be used as both a reductant and stabilizer. Scheme 1 shows the formula of OH-terminated PVP and the structural changes effected by $PtCl_6^{2-}$ ions. The powder X-Ray diffraction (XRD) pattern of the product (Figure 2) was matched with that of bulk Pt (Fm-3m, a = 3.924, PDF No. 89–7382), confirming the formation of metallic Pt USNPs with a face-centered cubic structure.

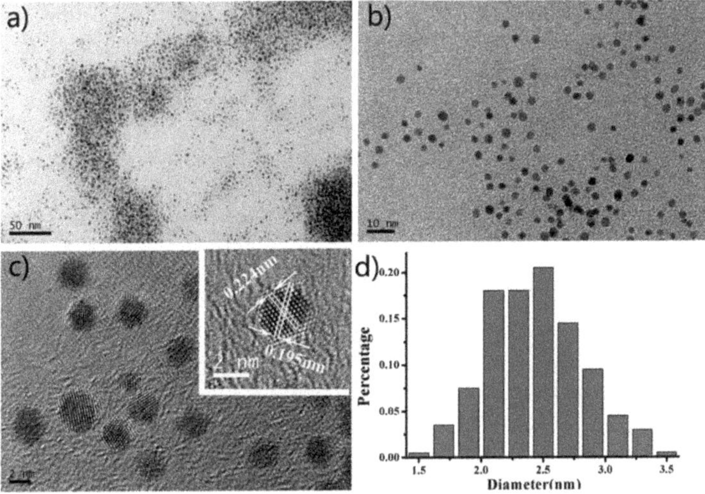

Figure 1. (**a**,**b**) Typical transmission electron microscope (TEM) images of ultrasmall Pt nanoparticles at different magnification and (**c**,**d**) high-resolution TEM image and the corresponding size distributions of the samples.

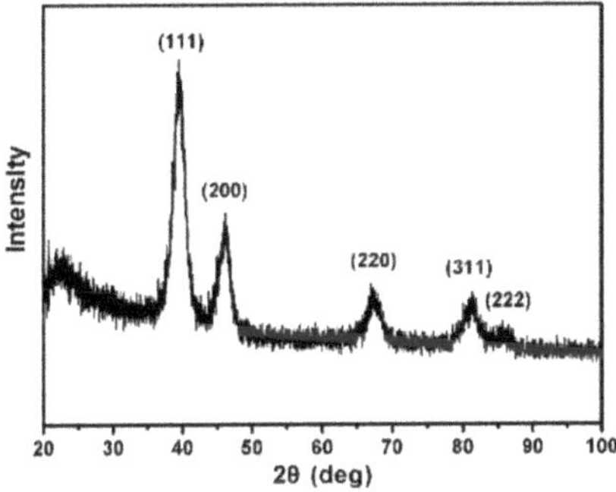

Figure 2. Powder X-Ray diffraction (XRD) pattern of ultrasmall Pt nanoparticles.

We tested the electrochemical behaviors of the prepared Pt USNPs. In order to minimize possible aggregation during the catalytic reaction and utilize the support effect, we used Vulcan XC-72 as the support. Vulcan XC-72 carbon, which is the most widely used support material for direct methanol fuel cell electrodes, is amorphous with a specific surface area of 212 $m^2{\cdot}g^{-1}$. A given weight of Vulcan XC-72 was mixed with the as-prepared solution of Pt USNPs for an hour. The products were separated and dried in a vacuum oven. Figure 3a shows that a number of Pt USNPs were distributed almost evenly over the entire surfaces of the Vulcan carbon powders. We measured cyclic voltammetry (CV) curves of Pt/Vulcan XC-72 catalysts in N_2-saturated 0.5 M H_2SO_4 in comparison with the commercial Pt/C catalyst (E-TEK, 30% Pt on Vulcan XC-72 carbon support). CV curves of Pt/Vulcan XC-72 catalysts (Figure 3b) reveal a large peak between −0.24 and 0.08 V (*vs.* saturated calomel electrode (SCE)), which is associated with the hydrogen adsorption/desorption processes. The electrochemically active surface area (ECSA) can be calculated from the charge involved in the hydrogen adsorption/desorption processes using Equation (1) [16].

$$ECSA = Q/[Pt] \times 0.21\ \text{mC}{\cdot}\text{cm}^{-2} \quad (1)$$

where Q (mC) and [Pt] are the charge for hydrogen adsorption/desorption and the loading of Pt on the electrode, respectively, while 0.21 $mC \cdot cm^{-2}$ is the electrical charge associated with the monolayer adsorption of hydrogen on Pt [17]. The ECSA of the Pt/Vulcan XC-72 is as high as 46.2 m^2/g_{Pt}, which is 1.52 times greater than that of the commercial Pt/C catalyst (30.3 m^2/g_{Pt}). It demonstrated Pt/Vulcan XC-72 catalysts had much higher ECSA.

The electrochemical performance of the Pt/Vulcan XC-72 composite (shown in Figure 3a) was tested for methanol oxidation, which is at the heart of direct methanol fuel cells (DMFC) application in the anodic half-cell reaction. For comparison, the commercial Pt/C catalyst was also evaluated in the same condition. The electrocatalytic performances of the Pt-based electrocatalysts towards methanol oxidation were evaluated using CV and amperometric *i-t* in 0.5 M H_2SO_4 + 0.5 M CH_3OH solution. CV curves in Figure 3c show two prominent peaks in both positive and negative scans corresponding to the methanol oxidation for all tested catalysts, and it can be seen that the current from methanol oxidation becomes apparent as the potential rises to above 0.2 V. In the forward scan, methanol oxidation produces an anodic peak at around 0.6 V. In the reverse scan, an anodic peak appears at around 0.4 V. This anodic peak in the reverse scan can be attributed to the removal of the incompletely oxidized carbonaceous species formed in the forward scan. These features of CV curves are in agreement with the reports for the Pt/C catalyst. As shown in Figure 3c, the peak current of methanol oxidation on Pt/Vulcan XC-72 is 373 $mA \cdot mg^{-1}$ Pt, which is much higher than that of the commercial Pt/C catalyst (148.3 $mA \cdot mg^{-1}$ Pt). The ratio of the forward oxidation current peak (I_f) to the reverse current peak (I_b), I_f/I_b, is an index of the catalyst tolerance to the poisoning species. A higher ratio indicates more effective removal of the poisoning species on the catalyst surface. The I_f/I_b ratio of Pt/Vulcan XC-72 is 2.27, higher than that of the commercial Pt/C (1.93), showing better catalyst tolerance of the Pt/Vulcan XC-72 catalyst.

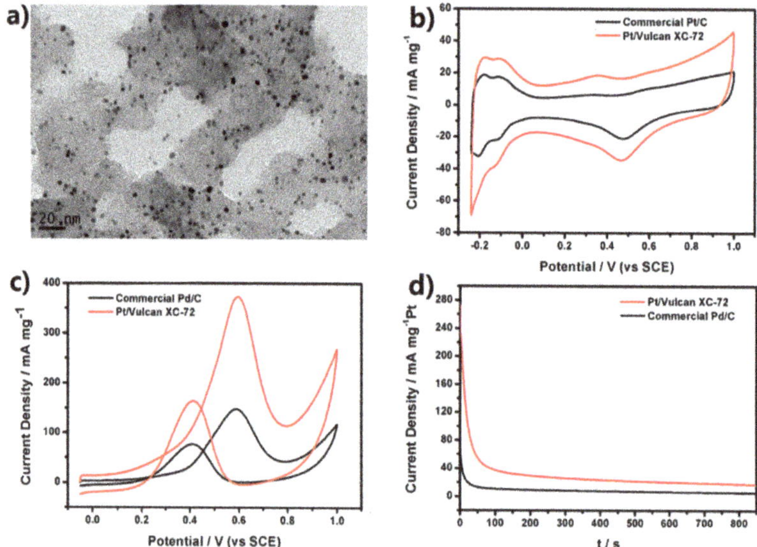

Figure 3. (**a**) TEM images of Pt ultrasmall nanoparticles (USNPs) supported on Vulcan XC-72 carbon; (**b**) Cyclic voltammetry (CV) curves of Pt electrocatalysts in N_2-saturated 0.5 M H_2SO_4 with a scan rate of 50 $mV \cdot s^{-1}$; (**c**) CV curves at a scan rate of 50 $mV \cdot s^{-1}$ with different Pt electrocatalysts in 0.5 M H_2SO_4 + 0.5 M CH_3OH; (**d**) Amperometric *i-t* curves of Pt electrocatalysts in 0.5 M H_2SO_4 + 0.5 M CH_3OH solution at 0.6 V (*vs.* saturated calomel electrode (SCE)).

To further compare the electrochemical long-term performance between Pt/Vulcan XC-72 and commercial Pt/C, a chronoamperometry test was conducted in the solution of 0.5 M H_2SO_4 and 0.5 M CH_3OH for 850 s at a fixed potential of 0.6 V (*vs.* SCE), as shown in Figure 3d. The initial high current density was attributed to the double-layer charging and the numerous available active sites of Pt atoms. The currents of all the catalysts decreased rapidly within first few seconds due to the formation of CO-like intermediates such as CO and CHO, *etc.* [18–21], which generated during the oxidation of methanol and adsorbed on the surface of active Pt atoms to prevent methanol's further oxidation on these Pt atoms. After that, the currents kept on decreasing and gradually maintained a steady state. The current density of Pt/Vulcan XC-72 was higher than that of commercial Pt/C during the whole testing time. The current density of Pt/Vulcan XC-72 is 17.2 $mA \cdot mg^{-1}$ at 850 s, which is much higher than that of commercial Pt/C (5.6 $mA \cdot mg^{-1}$ at 850 s). The higher current density directly reflected the higher resistivity of Pt/Vulcan XC-72 to the poisoning of CO-like intermediates.

3. Experimental Section

3.1. Chemicals and Materials

Potassium hexachloroplatinate (IV) (K_2PtCl_6, AR) and methanol were purchased from Shanghai Chemical Reagent Company (Shanghai, China). Poly(vinyl pyrrolidone) (PVP, M_w = 55,000) was received from Aldrich. Vulcan XC-72 was purchased from Cabot (Boston, MA, USA) and 5% Nafion was purchased from Sigma (Shanghai, China). All chemicals were used as received without further purification. High-purity deionized water (>18.4 $M\Omega \cdot cm$) was produced using Millipore A10 Milli-Q (Darmstadt, Germany).

3.2. Synthesis of Pt USNPs

A typical synthesis process can be concisely described as follows. K_2PtCl_6 (38.8 mg) and PVP (44.4 mg, M_w = 55,000) were added into water (10 mL), and the solution was stirred for 10 min. Then the homogeneous orange-yellow solution was transferred to a 15 mL Teflon-lined stainless steel autoclave. The container was then sealed in a stainless steel bomb. The whole system was heated and maintained in an oven at 160 °C under autogenous pressure for 8 h. After the reaction finished, the container was cooled under room temperature conditions naturally. Finally, a solution of Pt particles with dark gray color was obtained.

3.3. Synthesis of Pt USNPs/Vulcan XC-72

A given weight of Vulcan XC-72 was mixed with the as-prepared solution of Pt USNPs for an hour. The products were separated via centrifugation (12,000 rpm, 10 min) and further purified twice by deionized water and then dried in a vacuum oven at 80 °C for 8 h.

3.4. Characterization

The transmission electron microscopy (TEM) images were obtained on JEOL JEM-2100F instruments (Tokyo, Japan) operating at an accelerating voltage of 200 kV. For the preparation of samples for electron microscopy analyses, aliquots taken from the as-prepared Pt USNPs were dropped directly onto carbon-coated copper grids placed on a filter paper and then dried at room temperature in air. The grids were placed in a sack and steeped in alcohol for a half hour. After that, the grids were taken out and dried at room temperature in air. The structures of as-prepared products were characterized by X-Ray diffraction (XRD) (Rigaku D/Max 2550 X, Cu Kα radiation, λ = 0.154178 nm, Tokyo, Japan). The composition of Pt/Vulcan XC-72 was determined by energy dispersive X-Ray spectroscopy (EDS, Quanta FEG 250, 20 kV, FEI Company, Hillsboro, OR, USA).

3.5. Electrocatalytic Activity Evaluation

Electrochemical characterizations were carried out using with CHI660D and a three-electrode configuration with a saturated calomel electrode (SCE) and a platinum foil as the reference and counter electrode, respectively. The catalyst ink, which was prepared from the mixture of catalyst (2 mg), ethanol (1 mL), and Nafion solution (5%, 50 μL), was spread onto a glass carbon (GC) electrode (4 mm diameter). For all electrodes, a metal loading of 23.8 $\mu g \cdot cm^{-2}$ was used. Commercial Pt/C catalyst (30 wt %) was also tested for comparison. Cyclic voltammetry (CV) measurements were performed in 0.5 M H_2SO_4 solutions under a flow of N_2 at a sweep rate of 50 $mV \cdot s^{-1}$. The amperometric *i-t* curves were obtained in 0.5 M H_2SO_4 + 0.5 M CH_3OH.

4. Conclusions

In summary, a novel method has been successfully developed to synthesize Pt USNPs with an average size of 2.45 nm. These Pt USNPs supported on Vulcan XC-72 exhibited an enhanced activity for methanol oxidation compared to the commercial Pt/C catalyst, thus providing great potential as a promising electrocatalyst for high performance DMFCs. Our results suggest that the PVP-assisted hydrothermal method is a simple and environmentally benign route for the synthesis of Pt nanoparticles for catalysis and other applications. Significantly, our results also provide a new insight into the role played by the hydrothermal method in the synthesis of noble metal nanoparticles.

Acknowledgments: This work was supported by National Nature Science Foundation of China (No. 21373273), Hunan Provincial Natural Science Foundation of China (No. 13JJ1002), Fundamental Research Funds for the Central Universities of Central South University (No. 2014zzts168) and Shenghua Scholar Program of Central South University.

Author Contributions: Wenhai Ji performed the experimental and sample preparation, general data mining and analysis, and prepared the manuscript. Shasha Tang, Hongcheng Peng and Siqi Li performed the electrochemical characterizations and data analysis. This project was proposed by Weihong Qi, who also supervised the research process, discussed the experimental results, revised the manuscript and approved the final version.

Conflicts of Interest: The authors declare no conflict of interest.

References

1. Mourdikoudis, S.; Chirea, M.; Altantzis, T.; Pastoriza-Santos, I.; Pérez-Juste, J.; Silva, F.; Bals S. Liz-Marzán, L.M. Dimethylformamide-mediated synthesis of water-soluble platinum nanodendrites for ethanol oxidation electrocatalysis. *Nanoscale* **2013**, *5*, 4776–4784. [PubMed]
2. Peng, Z.; Kisielowski, C.; Bell, A.T. Surfactant-free preparation of supported cubic platinum nanoparticles. *Chem. Commun.* **2012**, *48*, 1854–1856.
3. Tian, N.; Zhou, Z.Y.; Sun, S.G.; Ding, Y.; Wang, Z.L. Synthesis of tetrahexahedral platinum nanocrystals with high-index facets and high electro-oxidation activity. *Science* **2007**, *316*, 732–735. [PubMed]
4. Chen, A.; Holt-Hindle, P. Platinum-based nanostructured materials: Synthesis, properties, and applications. *Chem. Rev.* **2010**, *110*, 3767–3804. [PubMed]
5. Matoh, L.; Škofic, I.K.; Čeh, M.; Bukovec, N. A novel method for preparation of a platinum catalyst at low temperatures. *J. Mater. Chem. A* **2013**, *1*, 1065–1069. [CrossRef]
6. Teranishi, T.; Hosoe, M.; Tanaka, T.; Miyake, M. Size control of monodispersed Pt nanoparticles and their 2D organization by electrophoretic deposition. *J. Phys. Chem. B* **1999**, *103*, 3818–3827. [CrossRef]
7. Giuffrida, S.; Ventimiglia, G.; Callari, F.L.; Sortino, S. Straightforward Light-Driven Synthesis of Ultrasmall, Water-Soluble Monolayer-Protected Platinum Nanoparticles. *Eur. J. Inorg. Chem.* **2006**, *20*, 4022–4025. [CrossRef]
8. Li, Y.; Whyburn, G.P.; Huang, Y. Specific peptide regulated synthesis of ultrasmall platinum nanocrystals. *J. Am. Chem. Soc.* **2009**, *131*, 15998–15999. [CrossRef] [PubMed]
9. Pelzer, K.; Hävecker, M.; Boualleg, M.; Candy, J.P.; Basset, J.M. Stabilization of 200-Atom Platinum Nanoparticles by Organosilane Fragments. *Angew. Chem.* **2011**, *123*, 5276–5279. [CrossRef]
10. Lim, G.H.; Yu, T.; Koh, T.; Lee, J.H.; Jeong, U.; Lim, B. Reduction by water for eco-friendly, capping agent-free synthesis of ultrasmall platinum nanocrystals. *Chem. Phys. Lett.* **2014**, *595*, 77–82. [CrossRef]

11. Maji, T.; Banerjee, S.; Biswas, M.; Mandal, T.K. *In situ* synthesis of ultra-small platinum nanoparticles using a water soluble polyphenolic polymer with high catalytic activity. *RSC Adv.* **2014**, *4*, 51745–51753. [CrossRef]
12. Xiong, Y.; Washio, I.; Chen, J.; Cai, H.; Li, Z.Y.; Xia, Y. Poly(vinyl pyrrolidone): A dual functional reductant and stabilizer for the facile synthesis of noble metal nanoplates in aqueous solutions. *Langmuir* **2006**, *22*, 8563–8570. [CrossRef] [PubMed]
13. Washio, I.; Xiong, Y.; Yin, Y.; Xia, Y. Reduction by the end groups of poly(vinyl pyrrolidone): A new and versatile route to the kinetically controlled synthesis of Ag triangular nanoplates. *Adv. Mater.* **2006**, *18*, 1745–1749. [CrossRef]
14. Ji, W.; Qi, W.; Tang, S.; Huang, B.; Wang, M.; Li, Y.; Jia, Y.; Pang, Y. Synthesis of Marks-Decahedral Pd Nanoparticles in Aqueous Solutions. *Part. Part. Syst. Charact.* **2014**, *31*, 851–856. [CrossRef]
15. Ji, W.; Qi, W.; Li, X.; Zhao, S.; Tang, S.; Peng, H.; Li, S. Investigation of disclinations in Marks decahedral Pd nanoparticles by aberration-corrected HRTEM. *Mater. Lett.* **2015**, *152*, 283–286. [CrossRef]
16. Pozio, A.; de Francesco, M.; Cemmi, A.; Cardellini, F.; Giorgi, L. Comparison of high surface Pt/C catalysts by cyclic voltammetry. *J. Power Sources* **2002**, *105*, 13–19. [CrossRef]
17. Maillard, F.; Martin, M.; Gloaguen, F.; Léger, J.M. Oxygen electroreduction on carbon-supported platinum catalysts. Particle-size effect on the tolerance to methanol competition. *Electrochim. Acta* **2002**, *47*, 3431–3440. [CrossRef]
18. Lee, S.H.; Kakati, N.; Jee, S.H.; Maiti, J.; Yoon, Y.S. Hydrothermal synthesis of PtRu nanoparticles supported on graphene sheets for methanol oxidation in direct methanol fuel cell. *Mater. Lett.* **2011**, *65*, 3281–3284. [CrossRef]
19. Prabhuram, J.; Zhao, T.S.; Tang, Z.K.; Chen, R.; Liang, Z.X. Multiwalled carbon nanotube supported PtRu for the anode of direct methanol fuel cells. *J. Phys. Chem. B* **2006**, *110*, 5245–5252. [CrossRef] [PubMed]
20. Zhang, Y.; Gu, Y.E.; Lin, S.; Wei, J.; Wang, Z.; Wang, C.; Du, Y.; Ye, W. One-step synthesis of PtPdAu ternary alloy nanoparticles on graphene with superior methanol electrooxidation activity. *Electrochim. Acta* **2011**, *56*, 8746–8751. [CrossRef]
21. Liu, F.; Lee, J.Y.; Zhou, W.J. Segmented Pt/Ru, Pt/Ni, and Pt/RuNi nanorods as model bifunctional catalysts for methanol oxidation. *Small* **2006**, *2*, 121–128. [CrossRef] [PubMed]

 *nanomaterials*

Communication

N-doped TiO_2 Nanotubes as an Effective Additive to Improve the Catalytic Capability of Methanol Oxidation for Pt/Graphene Nanocomposites

Xiaohua Wang, Yueming Li *, Shimin Liu and Long Zhang

State Key Laboratory of Metastable Materials Science and Technology, College of Materials Science and Engineering, Yanshan University, Qinhuangdao 066004, China; wangxiaohua2016@hotmail.com (X.W.); lsm@ysu.edu.cn (S.L.); lzhang@ysu.edu.cn (L.Z.)

* Correspondence: liyueming@ysu.edu.cn; Tel.: +86-335-8074-631

Academic Editors: Hermenegildo García and Sergio Navalón

Received: 11 December 2015; Accepted: 16 February 2016; Published: 26 February 2016

Abstract: *N*-doped TiO_2 nanotubes have been prepared as additives to improve the catalytic capability of Pt/graphene composites in methanol oxidation reactions. Electrochemical experiments show that the catalytic performance of Pt/graphene composites has been greatly improved by the introduction of *N*-doped TiO_2 nanotubes.

Keywords: methanol oxidation; TiO_2 nanotubes; doping; graphene

---

1. Introduction

Carbon materials are widely used as electrode materials in supercapacitors, batteries and catalyst carrier in fuel cells [1–3]. Novel carbon materials especially such as carbon nanotubes (CNTs), mesoporous carbon, carbon nanoparticles and graphene nanosheets have drawn much more attention in the field of energy storage and conversion due to their high conductivity, large surface area and good chemical stability [4,5].

The direct methanol fuel cell (DMFC) is considered a promising power source for portable electronic devices [6,7]. One key challenge is to develop catalysts with high catalytic capability. The form of carbon is found to play an important role for the catalytic performance of anode catalysts in DMFC [8]. Materials with high surface area will show obvious advantages due to the facilitated mass and charge transport and a higher electrode/electrolyte contact area [9]. Graphene nanosheets have shown improved catalytic performance as support carrier compared with traditional carbon black due to the high specific surface area and high conductivity, according to previous work by us and another group [10–12]. However, the catalytic performance for Pt/graphene based catalyst need to be further improved to reduce the usage of noble metal. A lot of methods have been used to improve the electrochemical performance of Pt based catalyst by modifying the support carrier. For example, nitrogen doping has proved to be an effective strategy to improve the electrochemical performance of graphene materials. A series of *N*-doped graphene/Pt nanocomposites have been prepared and showed improved electrochemical activity toward methanol oxidation [13–15].

Transition metal oxides such as CeO_2, SnO_2, MnO_2, and TiO_2 can be employed as support carriers to improve the electrocatalysts activity and stability [16,17]. Among these metal oxides, TiO_2 is of particular interest due to the good corrosion resistance, low price and environmentally-friendly nature. TiO_2 as a semiconductor has a low electric conductivity, which limited its total substitution of carbon support carriers. Considering the high conductivity of carbon materials, TiO_2/carbon materials composites are frequently used as support to overcome the disadvantage of TiO_2 [18,19].

Furthermore, the electrochemical performance of TiO_2 is also dependent on its crystal phase and chemical doping [20].

Considering the special properties of *N*-doped TiO_2, in this report, high performance Pt/graphene catalyst were prepared in which *N*-doped TiO_2 nanotubes was added as additives, and the catalytic performance toward methanol oxidation reaction indicated that as formed composites had much improved catalytic activity compared with Pt/graphene composites without additives.

2. Results and Discussion

Figure 1A shows the X-ray diffraction (XRD) pattern of TiO_2 nanotubes and *N*-doped TiO_2 nanotubes. For TiO_2 nanotubes, most of the strong peaks can be easily indexed to anatase-phase TiO_2 (JCPDS No. 21-1272). In contrast, the broadened peaks at ~26° for *N*-doped sample might be related to distortion in the O–Ti–O lattice due to the doping of nitrogen into TiO_2, similar to our previous report [21]. Figure 1B shows the N_2 adsorption/desorption isotherm at 77 K. These isotherms exhibit characteristics of characteristics of type IV, indicating the presence of meso and micro pores. The total specific surface areas (SSA) of *N*-doped and undoped TiO_2 evaluated using the Brunauer–Emmett–Teller (BET) equation were 257.7 and 267.6 $m^2 \cdot g^{-1}$, respectively. The *N*-doped TiO_2 sample showed a light yellow color, which is obviously different to the white color of undoped sample (inset of Figure 1B). As shown in Figure 1C, the *N*-doped TiO_2 nanotubes showed the typical hollow tubular morphology with a diameter ~7–10 nm, up to hundreds of nanometer in length, and the wall thickness of the nanotube is approximately 2–3 nm. Notably, TiO_2 nanorods may form under certain heat treatment of titanium hydrogen oxide nanotubes such as under hydrogen atmosphere [22]. The XRD patterns of the graphite oxide (GO), Pt/graphene (Pt/G), Pt/graphene with *N*-doped TiO_2 nanotubes (Pt/G-NT) and Pt/Graphene with undoped TiO_2 nanotubes (Pt/G-T) are shown in Figure 1D. As shown in the XRD pattern of GO, the characteristic peak at 10.8° indicated the successful oxidation of graphite. The strong diffraction peaks at ~39.9° and 46.2° in the XRD patterns correspond to the (111), and (200) facets of the face-centered cubic structures of platinum crystal, which are in good agreement with the cubic Pt (JCPDS No. 4-802). The broad peak located at ~24° indicates the reduction of graphite oxide into poorly ordered graphene nanosheets. Although *N*-doped TiO_2 nanotubes were added during the preparation process, no peaks corresponding to TiO_2 were detected due to the small amounts added.

The typical TEM image of Pt/G-NT nanocomposites (Figure 1D) showed that the Pt nanoparticles were deposited on the surface of graphene nanosheets. No TiO_2 nanotubes additive in the Pt/graphene composites can be seen from the TEM image due to the low ratio of GO: *N*-TiO_2 nanotubes and the reaction of *N*-TiO_2 in the solution. As shown in Figure 1E, it is clearly visible that Pt nanoparticles with diameter in the range of 3–10 nm were decorated on the surface of graphene nanosheets. High resolution transmission electron microscopy (HRTEM) investigation (Figure 1F) showed that the lattice fringes with a spacing of ~0.23 nm can be seen in these domains, consistent with the spacing of the (111) planes of Pt calculated from XRD.

The X-ray photoelectron spectroscopy (XPS) survey spectrum of as prepared *N*-doped TiO_2 (Figure 2A) confirms the presence of Ti, O, and N, while that of Pt/G-NT proves that the presence of Pt, C, O, Ti and N. The Ti 2p XPS spectrum (Figure 2B) can be deconvoluted into three peaks centered at about 284.8, 285.5 and 287.2 eV, corresponding to C–C, C–O and C=O bonding, respectively. Figure 2B illustrates the high-resolution XPS spectra of Ti 2p. Two prominent peaks located at about 458.7 and 464.2 eV for *N*-doped TiO_2 can be assigned to Ti 2p3/2 and Ti 2p1/2, respectively. The binging energy of Ti 2p peak shifts to lower energies compared to that of Ti^{4+}, indicating the presence of trivalent Ti bonding due to the partial substitution of O atom in the TiO_2 with a N atom [23]. In contrast, the doublet peaks for Pt/G-NT shifted to higher binding energies, suggesting the valence state of Ti^{3+} has changed to Ti^{4+}. The entire process is still not very clear. The possible process might be as follows. The "TiN" on the surface of *N*-doped TiO_2 nanotubes can react with water or NaOH in the solution [24]. During this process, trivalent Ti changed into tetravalent Ti^{4+}, while N was released in the form of NH_3.

The as-formed NH_3 in aqueous solution will react with GO, resulting to the formation of *N*-doped graphene. Quantitative XPS analysis shows that the doping level of Ti in Pt/G-NT is ~0.17 atomic (at.) %. The O 1s spectrum of *N*-doped TiO_2 is shown in Figure 2C with their peak curve-fitting lines with respect to the chemical states. The O 1s peaks can be fitted into two peaks at about 530.1 and 532.2 eV, which are assigned to Ti–O and chemisorbed oxygen, respectively. In comparison to the C 1s spectrum of the GO, that of Pt/G-NT sample showed sharply decreased intensity for peak(s) corresponding to the epoxy/ether group (286.9 eV, Figure 2D), indicating that most these oxygen containing groups have been removed during reaction. Notably, sp2 C–N peak is overlapping with C–O peak, and sp3 C–N peak is also overlapping with C=O peak [25–27]. The high-resolution N 1s XPS spectrum shown in Figure 2E reveals the presence of Ti-N bond as well as O−Ti−N bond (in as prepared *N*-doped TiO_2 nanotube. The nitrogen content is ~6.47 at % in the *N*-doped TiO_2 according to quantitative analysis. For Pt/G-NT sample, N 1s spectrum can also be observed. The results indicate that the small amount of N atoms is incorporated into the carbon framework of graphene sheets. Quantitative XPS analysis shows that the doping level of N in Pt/G-NT is ~0.86 at %. As seen from the high resolution N 1s spectrum (Figure 2D), the N 1s peak can be fitted by three component peaks at 398.3, 400.2 and 401.9 eV which can be attributed to pyridinic (N-6), pyrrolic/pyridine (N-5) and quaternary nitrogen (N-Q), respectively. In Figure 2F, the main doublet at 75.1 and 71.7 eV is characteristic of metallic Pt, indicating the reduction of tetravalent Pt.

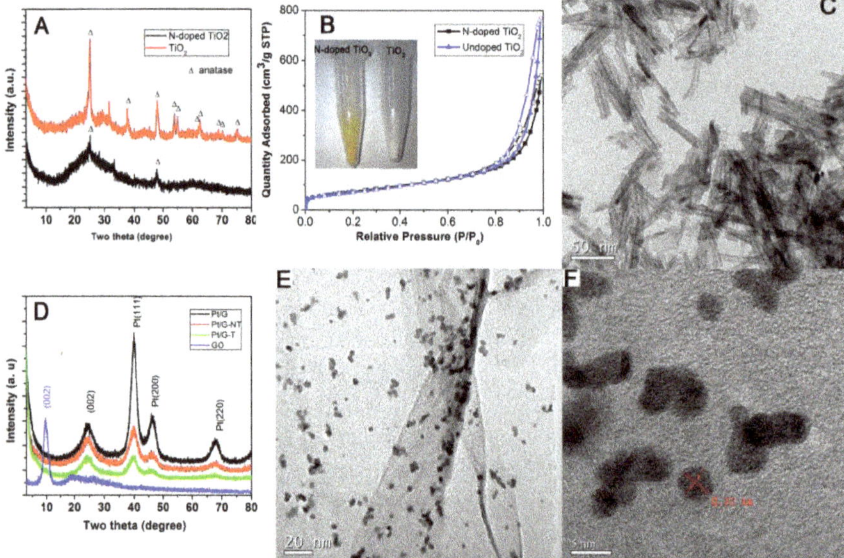

Figure 1. (**A**) X-ray diffraction (XRD) patterns of TiO_2 nanotubes and *N*-doped TiO_2 nanotubes; (**B**) N_2 adsorption/desorption isotherm for *N*-doped TiO_2 nanotubes and undoped TiO_2 at 77 K (inset: digital image of *N*-doped TiO_2 nanotube and undoped TiO_2 nanotubes); (**C**) Transmission electron microscopy (TEM) images of *N*-doped TiO_2 nanotubes; (**D**) XRD patterns of graphene oxide (GO, Pt/graphene (Pt/G), Pt/graphene with TiO_2 (Pt/G-T), and Pt/graphene with *N*-doped TiO_2 (Pt/G-NT); (**E**) TEM image of Pt/G-NT and (**F**) High-resolution transmission electron microscopy (HRTEM) of Pt/G-NT composites.

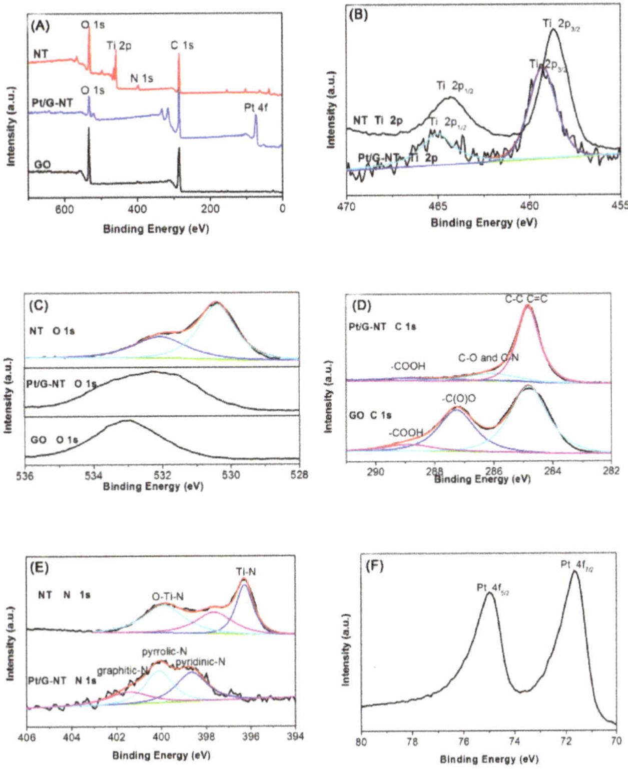

Figure 2. X-ray photoelectron survey spectra of GO, *N*-doped TiO_2 and Pt/G-NT (**A**), high resolution spectra of Ti 2p (**B**), O 1s (**C**), C 1s (**D**), N 1s (**E**) and Pt 4f (**F**).

To evaluate the electrochemical activity of as-prepared samples, cyclic voltammogram (CV) experiments were carried out within a potential range from −0.2 to 1.0 V at a scanning rate of 50 $mV\cdot s^{-1}$ in the solution of nitrogen saturated 0.5 M H_2SO_4. As seen in Figure 3, the Pt/G-NT electrode shows electrochemically active nature, an obvious hydrogen adsorption characteristic.

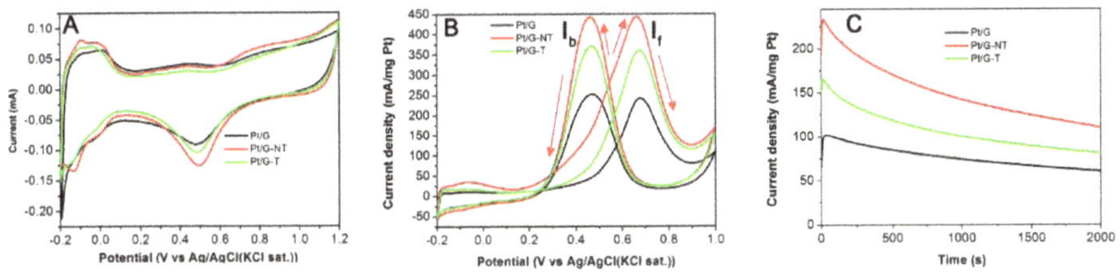

Figure 3. (**A**) Cyclic voltammograms of Pt/G, Pt/G-NT and Pt/G-T an in nitrogen saturated aqueous solution of 0.5 M H_2SO_4 at a scan rate of 50 $mV\cdot s^{-1}$. (**B**) Cyclic voltammograms of Pt/G, Pt/G-NT and Pt/G-T an in nitrogen saturated aqueous solution of 0.5 M H_2SO_4 containing 0.5 M CH_3OH at a scan rate of 50 $mV\cdot s^{-1}$. (**C**) Chronoamperometric curves for Pt/G, Pt/G-NT and Pt/G-T catalysts in nitrogen saturated aqueous solution of 0.5 M H_2SO_4 containing 0.5 M CH_3OH at a fixed potential of 0.5 V *vs.* Ag/AgCl (KCl saturated (sat.))

The electrochemically active surface areas (ECSAs) were evaluated by integrating the cyclic voltammogram corresponding to hydrogen desorption from the electrode surface [28]. The ECSAs for the Pt/G-NT, Pt/G and Pt/G-T were estimated to be 63, 50 and 58 $m^2 \cdot g^{-1}$ Pt, respectively. It is believed that the high ECSA helps to improve the electrochemical activity of the catalyst.

The electrochemical catalytic activity of the as prepared catalysts toward the methanol oxidation was evaluated by cyclic voltammetric experiments were tested in the solution of 0.5 M CH_3OH in 0.5 M H_2SO_4.

Figure 3 compares the electrochemical catalytic activities of Pt/G-NT, Pt/G-T and Pt/G. For the forward scan, the current (If) increased sharply attributed to the dehydrogenation of methanol and the following oxidation of the absorbed methanol on the electrode sites. The backward peak current (Ib) is related to the subsequent oxidation of the incompletely oxidized products during the forward scan. The ratio of the forward to backward peak current can be used to describe the tolerance of catalyst to the carbonaceous species accumulation [29,30]. As seen in the figure, the ratio of If/Ib for Pt/G-NT is slightly higher than those of Pt/G-T and Pt/G, suggesting that Pt/G-NT exhibits slightly better poisoning tolerance than Pt/G and Pt/G-T.

As seen from Figure 3A, the peak current of Pt/G-NT electrode was 446.8 mA/mg· Pt during the forward potential scanning process, which is much larger than those that for Pt/G-T and Pt/G (~360 and 240 mA/mg· Pt, respectively). Furthermore, the onset potential (where the forward peak current density begins to increase sharply in the CV curve) for Pt/G-NT electrode is also clearly lower than those of Pt/G-T and Pt/G, consistent with a previous report on the introduction of TiO_2 to Pt based catalyst [31]. Therefore, the performance of Pt/G-NT for the methanol electrochemical oxidation can be considered superior to that of Pt/G or Pt/G-T. As predicted by XPS, the presence of tetravalent titanium oxide is able to enhance dispersion of the wetting process due to the rich active –OH species, enlarging the electrode–electrolyte interfacial area and increasing the concentration of methanol confined around Pt catalyst [16,32–35]. Furthermore, the improved electrochemical performance is attributed to the *N*-doping effect on graphene nanosheets, because the nitrogen doping can intrinsically regulate the properties of carbon materials in modifying the electronic and chemical properties due to its comparable atomic size and five valence electrons available [36–38]. In addition, graphene sheets prepared via GO in the catalysts as an ideal support carrier can provide high surface area, anchor sites to Pt attributed to oxygen-containing groups as well as the high electronic conductivity. Furthermore, the homogeneous dispersion of Pt particles with nano sizes on the 2D graphene nanosheets can maximize the utilization of Pt.

The catalytic stability of the Pt/G, Pt/G-T and Pt/G-NT were examined using chronoamperometry. Figure 3C showed the chronoamperometric curves of 0.5 M CH_3OH in 0.5 M H_2SO_4 solution for these catalyst electrodes at a fixed potential of 0.50 V for 2000 s. It can be clearly observed that the potentiostatic current decreased rapidly at the initial stage for these three electrodes, which might be due to the formation of intermediate species, such as CO_{ads}, CHO_{ads}, *etc.*, during the methanol oxidation reaction [39]. It is obvious that Pt/G-NT retains the highest current density among these samples during the whole testing time, indicating that the electrocatalytic stability of the Pt/G-NT catalyst for the methanol oxidation was also higher than that of Pt/G-T or Pt/G. The doping of nitrogen into graphene will enhance the electrochemical performance of graphene. The improved electrochemical performance for Pt/G-NT can be attributed to the remaining TiO_2 and the nitrogen doping of graphene nanosheets.

3. Experimental Section

3.1. Method

Reagent grade chemicals were analytical purity unless otherwise stated. Graphene oxide (GO) was synthesized from graphite powder based on the modified Hummers method as described elsewhere [40]. Titanium hydrogen oxide nanotubes were prepared by hydrothermal method according to previous report [41,42]. To prepare *N*-doped TiO_2 nanotubes, as-prepared titanium hydrogen oxide nanotubes were heated in the presence of urea (TiO_2: Urea, 1:3 mass ratio) at 500 °C for 2 h under Ar atmosphere, yielding yellow powder. The as-prepared product was washed with deionized (DI) water and dried at 100 °C overnight. For comparison, undoped TiO_2 nanotubes were prepared with the similar condition except that no urea was added during heat treatment, yielding white powder.

To prepare *N*-doped graphene/Pt nanocomposites, GO (80 mg) and *N*-doped TiO_2 (2 mg) were dispersed in ethylene glycol (EG) (150 mL) and ultrasonic treated for 2 h to form the uniform dispersion solution. Chloroplatinic acid solution ($H_2PtCl_6 \cdot 6H_2O$, 5 wt % aqueous solution, 15 mL) was added to the dispersion. The pH of the mixture was adjusted to ~13 by NaOH aqueous solution. $NaBH_4$ (100 mg) was slowly added to the dispersion solution. The dispersion was heated at 140 °C for 4 h under magnetic stirring with Ar bubbling. The mixture was then filtered, washed with DI water and ethanol and then freeze-dried overnight, yielding black product, denoted as Pt/G-NT. Pt/graphene nanocomposites with the same Pt loading were prepared with the same procedure except undoped TiO_2 nanotubes were added, and the final product was denoted as Pt/G-T. For comparison, Pt/graphene with no TiO_2 added was also prepared via the same procedure, denoted as Pt/G. The loading of Pt for all catalysts is ~40 wt %.

3.2. *Characterization*

The powder X-ray diffraction (XRD) measurements of the samples were recorded on a Bruker D8-Advance X-ray powder diffractometer (Karlsruhe, Germany) using Cu Kα radiation (λ = 1.5406 Å) with scattering angles (2θ) of 10°–80°. A JEOL JEM 2010 transition electronic microscopy (Tokyo, Japan) was used for transmission electron microscopy (TEM) analysis and high-resolution transmission electron microscopy (HRTEM) analysis. The Brunauer–Emmett–Teller (BET) specific surface area was calculated from N_2 adsorption/desorption isotherms which were obtained by a gas adsorption analyzer (ASAP 2020, Micromeritics Instrument Co. Norcross, GA, USA) at 77 K. X-ray photoelectron spectroscopy (XPS) was carried out on ESCALAB 250XI (Waltham, MA, USA) and the binding energy is calibrated with C 1s = 284.8 eV.

3.3. *Electrochemical Measurements*

Electrochemical measurements were performed on a Princeton P4000 electrochemical working station (Oak Ridge, TN, USA) with a standard three-electrode electrochemical cell. The catalyst electrodes were prepared as follows: 1.0 mg catalyst in 1.0 mL ethanol with Nafion solution was ultrasonicated for 30 min. Then, 5 μL of this suspension was transferred onto a glassy carbon electrode (GC, 3 mm diameter), dried overnight, and used as the working electrode. An Ag/AgCl (saturated (sat.) KCl) electrode was used as the reference and a platinum foil was used as the counter electrode.

4. Conclusions

In conclusion, a Pt/graphene nanocomposites catalyst with improved electrochemical performance has been prepared via a facile solution synthesis procedure. The electrochemical experiment proved that the addition of *N*-doped TiO_2 nanotubes is able to significantly improve the catalytic performance of Pt/graphene composites. The peak current of Pt/G-NT was nearly twice that of unmodified Pt/G catalysts. The stability of *N*-doped TiO_2 modified catalyst was also much improved compared to unmodified Pt/graphene catalyst, indicating the *N*-doped TiO_2 nanotube is a very effective additive to modify the electrochemical performance of a Pt-based catalyst.

Acknowledgments: This work was supported by the National Natural Science Foundation of China (No. 50122212) and Natural Science Foundation of Hebei Province (Project No. E2014203033).

Conflicts of Interest: The authors declare no conflict of interest.

References

1. Girishkumar, G.; Vinodgopal, K.; Kamat, P.V. Carbon nanostructures in portable fuel cells: Single-walled carbon nanotube electrodes for methanol oxidation and oxygen reduction. *J. Phys. Chem. B* **2004**, *108*, 19960–19966. [CrossRef]
2. Pandolfo, A.; Hollenkamp, A. Carbon properties and their role in supercapacitors. *J. Power Sources* **2006**, *157*, 11–27. [CrossRef]
3. Zhang, L.L.; Zhao, X. Carbon–based materials as supercapacitor electrodes. *Chem. Soc. Rev.* **2009**, *38*, 2520–2531. [CrossRef] [PubMed]
4. Wang, Y.; Shi, Z.; Huang, Y.; Ma, Y.; Wang, C.; Chen, M.; Chen, Y. Supercapacitor devices based on graphene materials. *J. Phys. Chem. C* **2009**, *113*, 13103–13107. [CrossRef]
5. Simon, P.; Gogotsi, Y. Capacitive energy storage in nanostructured carbon-electrolyte systems. *Acc. Chem. Res.* **2012**, *46*, 1094–1103. [CrossRef] [PubMed]
6. Liu, H.; Song, C.; Zhang, L.; Zhang, J.; Wang, H.; Wilkinson, D.P. A review of anode catalysis in the direct methanol fuel cell. *J. Power Sources* **2006**, *155*, 95–110. [CrossRef]
7. Zhu, C.; Du, D.; Eychmüller, A.; Lin, Y. Engineering ordered and nonordered porous noble metal nanostructures: Synthesis, assembly, and their applications in electrochemistry. *Chem. Rev.* **2015**, *115*, 8896–8943. [CrossRef] [PubMed]
8. Li, W.; Liang, C.; Zhou, W.; Qiu, J.; Zhou, Z.; Sun, G.; Xin, Q. Preparation and characterization of multiwalled carbon nanotube–supported platinum for cathode catalysts of direct methanol fuel cells. *J. Phys. Chem. B* **2003**, *107*, 6292–6299. [CrossRef]
9. Arico, A.S.; Bruce, P.; Scrosati, B.; Tarascon, J.M.; van Schalkwijk, W. Nanostructured materials for advanced energy conversion and storage devices. *Nat. Mater.* **2005**, *4*, 366–377. [CrossRef] [PubMed]
10. Li, Y.; Tang, L.; Li, J. Preparation and electrochemical performance for methanol oxidation of Pt/graphene nanocomposites. *Electrochem. Commun.* **2009**, *11*, 846–849. [CrossRef]
11. Jang, H.D.; Kim, S.K.; Chang, H.; Choi, J.-H.; Cho, B.-G.; Jo, E.H.; Choi, J.-W.; Huang, J. Three-dimensional crumpled graphene-based platinum-gold alloy nanoparticle composites as superior electrocatalysts for direct methanol fuel cells. *Carbon* **2015**, *93*, 869–877. [CrossRef]
12. Chen, D.; Tang, L.H.; Li, J.H. Graphene-based materials in electrochemistry. *Chem. Soc. Rev.* **2010**, *39*, 3157–3180. [CrossRef]
13. Xiong, B.; Zhou, Y.; Zhao, Y.; Wang, J.; Chen, X.; O'Hayre, R.; Shao, Z. The use of nitrogen-doped graphene supporting Pt nanoparticles as a catalyst for methanol electrocatalytic oxidation. *Carbon* **2013**, *52*, 181–192. [CrossRef]
14. Xu, X.; Zhou, Y.K.; Lu, J.M.; Tian, X.H.; Zhu, H.X.; Liu, J.B. Single-step synthesis of PtRu/*N*-doped graphene for methanol electrocatalytic oxidation. *Electrochim. Acta* **2014**, *120*, 439–451. [CrossRef]
15. Zhao, S.L.; Yin, H.J.; Du, L.; Yin, G.P.; Tang, Z.Y.; Liu, S.Q. Three dimensional *N*-doped graphene/PtRu nanoparticle hybrids as high performance anode for direct methanol fuel cells. *J. Mater. Chem. A* **2014**, *2*, 3719–3724. [CrossRef]
16. Kakati, N.; Maiti, J.; Lee, S.H.; Jee, S.H.; Viswanathan, B.; Yoon, Y.S. Anode catalysts for direct methanol fuel cells in acidic media: Do we have any alternative for Pt or Pt–Ru? *Chem. Rev.* **2014**, *114*, 12397–12429. [CrossRef] [PubMed]
17. Yu, X.; Kuai, L.; Geng, B. CeO_2/rGO/Pt sandwich nanostructure: rGO–enhanced electron transmission between metal oxide and metal nanoparticles for anodic methanol oxidation of direct methanol fuel cells. *Nanoscale* **2012**, *4*, 5738–5743. [CrossRef] [PubMed]
18. Jiang, Z.-Z.; Gu, D.-M.; Wang, Z.-B.; Qu, W.-L.; Yin, G.-P.; Qian, K.-J. Effects of anatase TiO_2 with different particle sizes and contents on the stability of supported Pt catalysts. *J. Power Sources* **2011**, *196*, 8207–8215. [CrossRef]

19. Xia, B.Y.; Wu, H.B.; Chen, J.S.; Wang, Z.; Wang, X.; Lou, X.W. Formation of Pt-TiO_2-rGO 3-phase junctions with significantly enhanced electro-activity for methanol oxidation. *Phys. Chem. Chem. Phys.* **2012**, *14*, 473–476. [CrossRef] [PubMed]
20. Ramadoss, A.; Kim, S.J. Facile preparation and electrochemical characterization of graphene/ZnO nanocomposite for supercapacitor applications. *Mater. Chem. Phys.* **2013**, *140*, 405–411. [CrossRef]
21. Di Valentin, C.; Pacchioni, G.; Selloni, A.; Livraghi, S.; Giamello, E. Characterization of paramagnetic species in *N*-doped TiO_2 powders by EPR spectroscopy and DFT calculations. *J. Phys. Chem. B* **2005**, *109*, 11414–11419. [CrossRef] [PubMed]
22. Zheng, Z.; Huang, B.; Lu, J.; Wang, Z.; Qin, X.; Zhang, X.; Dai, Y.; Whangbo, M.-H. Hydrogenated titania: Synergy of surface modification and morphology improvement for enhanced photocatalytic activity. *Chem. Commun.* **2012**, *48*, 5733–5735. [CrossRef] [PubMed]
23. Chen, X.; Burda, C. Photoelectron spectroscopic investigation of nitrogen-doped titania nanoparticles. *J. Phys. Chem. B* **2004**, *108*, 15446–15449. [CrossRef]
24. Sakka, Y.; Ohno, S.; Uda, M. Oxidation and degradation of titanium nitride ultrafine powders exposed to air. *J. Am. Chem. Soc.* **1992**, *75*, 244–248. [CrossRef]
25. Kim, J.-G.; Shi, D.; Kong, K.-J.; Heo, Y.-U.; Kim, J.H.; Jo, M.R.; Lee, Y.C.; Kang, Y.-M.; Dou, S.X. Structurally and electronically designed TiO_2N_x nanofibers for lithium rechargeable batteries. *ACS Appl. Mater. Interfaces* **2013**, *5*, 691–696. [CrossRef] [PubMed]
26. Wu, Z.-S.; Winter, A.; Chen, L.; Sun, Y.; Turchanin, A.; Feng, X.; Müllen, K. Three-dimensional nitrogen and boron co-doped graphene for high-performance all-solid-state supercapacitors. *Adv. Mater.* **2012**, *24*, 5130–5135. [CrossRef] [PubMed]
27. Lin, Z.; Waller, G.; Liu, Y.; Liu, M.; Wong, C.P. Facile synthesis of nitrogen-doped graphene via pyrolysis of graphene oxide and urea, and its electrocatalytic activity toward the oxygen-reduction reaction. *Adv. Energy Mater.* **2012**, *2*, 884–888. [CrossRef]
28. Søgaard, M.; Odgaard, M.; Skou, E.M. An improved method for the determination of the electrochemical active area of porous composite platinum electrodes. *Solid State Ion.* **2001**, *145*, 31–35. [CrossRef]
29. Kang, Y.; Pyo, J.B.; Ye, X.; Gordon, T.R.; Murray, C.B. Synthesis, shape control, and methanol electro-oxidation properties of Pt–Zn alloy and Pt_3Zn intermetallic nanocrystals. *ACS Nano* **2012**, *6*, 5642–5647. [CrossRef] [PubMed]
30. Sanetuntikul, J.; Ketpang, K.; Shanmugam, S. Hierarchical nanostructured Pt_8Ti-TiO_2/C as an efficient and durable anode catalyst for direct methanol fuel cells. *ACS Catal.* **2015**, *5*, 7321–7327. [CrossRef]
31. Shanmugam, S.; Gedanken, A. Carbon-coated anatase TiO_2 nanocomposite as a high-performance electrocatalyst support. *Small* **2007**, *3*, 1189–1193. [CrossRef] [PubMed]
32. Tamizhmani, G.; Capuano, G.A. Improved electrocatalytic oxygen reduction performance of platinum ternary alloy-oxide in solid-polymer-electrolyte fuel cells. *J. Electrochem. Soc.* **1994**, *141*, 968–975. [CrossRef]
33. Zhu, J.; Zhao, X.; Xiao, M.; Liang, L.; Liu, C.; Liao, J.; Xing, W. The construction of nitrogen-doped graphitized carbon-TiO_2 composite to improve the electrocatalyst for methanol oxidation. *Carbon* **2014**, *72*, 114–124. [CrossRef]
34. Shanmugam, S.; Gedanken, A. Synthesis and electrochemical oxygen reduction of platinum nanoparticles supported on mesoporous TiO_2. *J. Phys. Chem. C* **2009**, *113*, 18707–18712. [CrossRef]
35. Tian, M.; Wu, G.; Chen, A. Unique electrochemical catalytic behavior of Pt nanoparticles deposited on TiO_2 nanotubes. *ACS Catal.* **2012**, *2*, 425–432. [CrossRef]
36. Gong, K.; Du, F.; Xia, Z.; Durstock, M.; Dai, L. Nitrogen-doped carbon nanotube arrays with high electrocatalytic activity for oxygen reduction. *Science* **2009**, *323*, 760–764. [CrossRef] [PubMed]
37. Geng, D.; Yang, S.; Zhang, Y.; Yang, J.; Liu, J.; Li, R.; Sham, T.-K.; Sun, X.; Ye, S.; Knights, S. Nitrogen doping effects on the structure of graphene. *Appl. Surf. Sci.* **2011**, *257*, 9193–9198. [CrossRef]
38. Zhu, J.; Xiao, M.; Zhao, X.; Li, K.; Liu, C.; Xing, W. Nitrogen-doped carbon-graphene composites enhance the electrocatalytic performance of the supported Pt catalysts for methanol oxidation. *Chem. Commun.* **2014**, *50*, 12201–12203. [CrossRef] [PubMed]
39. Kabbabi, A.; Faure, R.; Durand, R.; Beden, B.; Hahn, F.; Leger, J.M.; Lamy, C. *In situ* FTIRS study of the electrocatalytic oxidation of carbon monoxide and methanol at platinum-ruthenium bulk alloy electrodes. *J. Electroanal. Chem.* **1998**, *444*, 41–53. [CrossRef]

40. Hummers, W.S.; Offeman, R.E. Preparation of graphitic oxide. *J. Am. Chem. Soc.* **1958**, *80*, 1339. [CrossRef]
41. Li, Y.; Wang, Z.; Lv, X.-J. *N*-doped TiO_2 nanotubes/*N*-doped graphene nanosheets composites as high performance anode materials in lithium-ion battery. *J. Mater. Chem. A* **2014**, *2*, 15473–15479. [CrossRef]
42. Perera, S.D.; Mariano, R.G.; Vu, K.; Nour, N.; Seitz, O.; Chabal, Y.; Balkus, K.J., Jr. Hydrothermal synthesis of graphene-TiO_2 nanotube composites with enhanced photocatalytic activity. *ACS Catal.* **2012**, *2*, 949–956. [CrossRef]

Article

Investigation of MnO_2 and Ordered Mesoporous Carbon Composites as Electrocatalysts for Li-O_2 Battery Applications

Chih-Chun Chin, Hong-Kai Yang and Jenn-Shing Chen *

Department of Applied Chemistry, National University of Kaohsiung, Kaohsiung City 811, Taiwan; ccchin17@gmail.com (C.-C.C.); hong-kai84@hotmail.com (H.-K.Y.)

* Correspondence: jschen@nuk.edu.tw; Tel.: +886-7-591-9463; Fax: + 886-7-591-9348

Academic Editors: Hermenegildo García and Sergio Navalón

Received: 9 November 2015; Accepted: 12 January 2016; Published: 18 January 2016

Abstract: The electrocatalytic activities of the MnO_2/C composites are examined in Li-O_2 cells as the cathode catalysts. Hierarchically mesoporous carbon-supported manganese oxide (MnO_2/C) composites are prepared using a combination of soft template and hydrothermal methods. The composites are characterized by X-ray powder diffraction, scanning electron microscopy, transmission electron microscopy, small angle X-ray scattering, The Brunauer–Emmett–Teller (BET) measurements, galvanostatic charge-discharge methods, and rotating ring-disk electrode (RRDE) measurements. The electrochemical tests indicate that the MnO_2/C composites have excellent catalytic activity towards oxygen reduction reactions (ORRs) due to the larger surface area of ordered mesoporous carbon and higher catalytic activity of MnO_2. The O_2 solubility, diffusion rates of O_2 and $O_2^{\bullet -}$ coefficients (D_{O_2} and $D_{O_2^-}$), the rate constant (k_f) for producing $O_2^{\bullet -}$, and the propylene carbonate (PC)-electrolyte decomposition rate constant (k) of the MnO_2/C material were measured by RRDE experiments in the 0.1 M $TBAPF_6$/PC electrolyte. The values of k_f and k for MnO_2/C are 4.29×10^{-2} cm·s^{-1} and 2.6 s^{-1}, respectively. The results indicate that the MnO_2/C cathode catalyst has higher electrocatalytic activity for the first step of ORR to produce $O_2^{\bullet -}$ and achieves a faster PC-electrolyte decomposition rate.

Keywords: MnO_2/C; cathode; lithium-oxygen battery; rotating ring-disk electrode

---

1. Introduction

Energy storage devices with high energy and power densities are being developed for use as power sources for electric vehicles (EV) and hybrid electric vehicles (HEV) [1–3]. Over the past few decades, the vast majority of relevant research has focused on upgrading the performance of conventional lithium-ion batteries for EV or HEV applications; however, their energy densities and specific charge capacities still fail to satisfy commercial requirements such as long-range driving, low cost, and fast charging [1,2,4]. In recent years, rechargeable nonaqueous Li-air batteries have attracted much interest owing to their low cost, environmental friendliness, and high theoretical energy density (~3500 Wh·kg^{-1}), which is nearly equivalent to a nine-fold increase over conventional Li-ion batteries (~400 Wh·kg^{-1}) [4–7]. Despite these favorable characteristics, their practical applications are still hampered by several serious challenges including limited rate capability, poor cycling stability due to the instability of the electrode and electrolyte, and low round-trip efficiency induced by the rather large polarization, resulting in a wide charge–discharge voltage gap [3,8–15]. These critical problems are highly attributable to the O_2 cathode.

A typical rechargeable Li-O_2 battery is constituted by a porous oxygen diffusion cathode, a lithium metal anode, and an Li^+-conducting electrolyte. In general, the O_2 cathode is an oxygen catalyst loaded with porous carbon material, which enables both Li_2O_2 deposition (oxygen reduction reactions, ORRs) and decomposition (oxygen evolution reactions, OERs) reactions to occur upon battery discharge and charge, respectively. Many reports [1,4–6,8,9,11,15–18] have pointed out that the electrochemical performance of Li-O_2 batteries depends on many factors such as: the nature and microstructure of the O_2 electrode, electrolyte formula (especially, the composition of solvent), O_2 partial pressure, possible presence of reactive contaminants (e.g., trace water), and the choice of catalysts. In order to enhance the properties of rechargeable Li-O_2 batteries, several strategies have been followed over the years to explore the electrolyte formula, choice, and microstructure design of the O_2 electrode and optimization of the operating parameters [1,3,5,8–11].

Carbon materials with various nanostructures have been developed and used as O_2 cathodes in Li-O_2 batteries [4,6,10,19]. It has been well demonstrated that the performance of Li-O_2 batteries is related to the properties of carbon, such as the morphology, surface area, porous structure, and conductivity [6,9,20]. The design of porous carbon cathodes requires larger intraparticulate voids and open frameworks in their architecture structure to accommodate the insoluble discharge products. These voids and frameworks should help improve discharge capacity and cycling performance [19–21]. Obviously, designing an optimum pore structure for carbon materials can effectively improve the electrochemical performance of Li-O_2 batteries. Although various porous carbon structures have been explored, some studies have demonstrated that hierarchically porous honeycomb-like carbon cathodes with mesoporous/macroporous pore size can increase the specific capacity of Li-O_2 batteries [4–6,15,19–26]. Moreover, it is well known that an ideal cathode catalyst can facilitate the complete reversibility of ORRs and OERs with low polarization in Li-O_2 batteries [21]. Several potential catalysts have recently been proposed to promote ORRs and OERs, including nitrogen-doped carbon, metal oxides, metal nitrides, precious and nonprecious metals, *etc.* [1,3,8,13,15,19,27–29]. Among metal oxides, MnO_2 is a catalyst material of great interest because of its low cost, environmental friendliness, abundance, and electrocatalytic activity for ORRs in Li-O_2 batteries [13,28,30–32]. This study of Li-O_2 batteries focuses on MnO_2-based catalysts.

In the first part of this work, we created a hierarchically mesoporous carbon-supported β-manganese oxide (MnO_2/C) as an O_2 cathode material. We present a detailed study of the Li-O_2 electrochemistry of the MnO_2/C material using an electrolyte of 1 M $LiPF_6$ in a propylene carbonate (PC, which was used in many of the initial works on Li-O_2 batteries) solvent. Although there have been many studies of MnO_2/C materials for Li-O_2 battery applications, few studies have examined the poor stability of the electrolyte due to its reaction with the superoxide radical ($O_2^{\bullet-}$) produced upon the discharge at the MnO_2/C electrode. In this work, the stability of the electrolyte against the $O_2^{\bullet-}$ of the MnO_2/C electrode was first explored by the RRDE technique. The RRDE was developed about 50 years ago and has been verified to be a powerful tool for the study of electrochemical reactions. RRDE consists of two concentric electrodes (disk and ring electrodes) in a cylindrical holder with both of the electrodes facing downward into the solution. Products generated at the disk reaction are swept outward by the convection caused by rotation, and can be detected electrochemically at the ring by fixing the potential on the ring electrode. In this study, a disk electrode coated with MnO_2/C materials and a Pt ring electrode was fixed at an $O_2^{\bullet-}/O_2$ oxidation potential to collect the $O_2^{\bullet-}$ ions in electrolytes. Therefore, in the second part, we emphasize aspects of the PC-based electrolyte reaction against $O_2^{\bullet-}$ and the related kinetic information of $O_2^{\bullet-}$ in the MnO_2/C electrode by studying rotating ring disk electrode (RRDE) experiments and using a lithium-free non-aqueous electrolyte due to the stability of the intermediate $O_2^{\bullet-}$. In addition, the oxygen solubility in the electrolyte and the oxygen diffusion velocity throughout the whole O_2 electrode play key roles in determining battery performance, especially at high current densities [33]. In this work, the O_2 solubility, diffusion rates of O_2 and superoxide radical ($O_2^{\bullet-}$) coefficients (D_{O_2} and $D_{O_2^-}$), rate constant (k_f) for producing $O_2^{\bullet-}$, and PC-electrolyte decomposition rate constant (k) of the MnO_2/C electrode were quantified.

2. Experimental Methods

MnO_2/C composites were prepared by supramolecular self-assembly methods followed by a hydrothermal process. A modification of the mesoporous metal oxides and carbon nanocomposites procedure of Huang *et al.* [34] was applied to synthesize the MnO_2/C composites. The first step was to synthesize a 20 wt. % resol ethanolic solution according to an established method [34,35]. A solution was prepared by dissolving 1.5 g of triblock copolymer Pluronic F127 ($OH(CH_2CH_2O)_n$-$(CH_2CH(CH_3)O)_m$-$(CH_2CH_2O)_nH$, $EO_{106}PO_{70}EO_{106}$, Sigma Aldrich, St. Louis, MO, USA) in 10 g of anhydrous ethanol, then 5 g 20 wt. % resol ethanolic solution and 0.28 g $MnCl_2 \cdot 4H_2O$ (J.T. Baker, 99.8%) were added into the above solution slowly under stirring for 30 min at an ambient temperature. The homogeneous mixture was then transferred into a Petri dish at an ambient temperature for 24 h. After being dried, the films were heated at 100 °C for another 24 h to form orange transparent membranes. The as-made products were scraped from the Petri dish and ground into powders and then calcinated at 400 °C for 5 h under an Ar atmosphere with a heating rate of 1 $°C \cdot min^{-1}$ to yield Mn/C powders. To obtain MnO_2/C composites, the as-prepared Mn/C powders were subjected to a hydrothermal process at 180 °C for 12 h with 0.22 g $KMnO_4$ (J.T. Baker) and 30 mL of deionized water in a Teflon-lined stainless steel autoclave.

A Rigaku-D/MaX-2550 diffractometer (Rigaku, Tokyo, Japan) with Cu K_α radiation (λ = 1.54 Å) was used to obtain X-ray diffraction (XRD) patterns for the samples. Small angle X-ray scattering (SAXS) measurements were taken on a Nanostar U small-angle X-ray scattering system (Bruker, Karlsruhe, Germany) using Cu K_α radiation (40 kV, 35 mA). The morphology of the sample was observed using a scanning electron microscope (SEM, Hitachi S-3400 (Hitachi Limited, Tokyo, Japan)) and transmission electron microscope (TEM, JEOL JEM-3010 (JEOL, Tokyo, Japan)). Selected area electron diffraction (SAED) was applied to examine samples' crystallinity. The Brunauer–Emmett–Teller (BET) method was used to measure the specific surface area of the powders (ASAP2020). The residual carbon content of the samples was measured by an automatic elemental analyzer (EA, Elementar vario, EL III (Elementar Analysensysteme GmbH, Hanau, Germany)).

For electrochemical evaluation, the MnO_2/C electrodes were prepared by wet coating, and were made from as-prepared MnO_2/C composites with super P and a poly(vinylidene difluoride) (PVDF) binder (MKB-212C, Atofina, Serquigny, France) in a weight ratio of 64:16:20. The MnO_2/C composites and super-P were first added to a solution of PVDF in *N*-methyl-2-pyrrolidone (NMP, Riedel-deHaen, Seelze, Germany). To make a slurry with an appropriate viscosity, the mixture was stirred for 20 min at room temperature using a magnetic bar, and then for 5 min using a turbine at 2000 rpm. The resulting slurry was coated onto a piece of separator (Celgard 2400, Charlotte, NC, USA) and dried at 60 °C under vacuum for 12 h. The coating had a thickness of ~100 μm with an active material mass loading of 8 $\pm$ 1 $mg \cdot cm^{-2}$. The quantity of active materials on the electrodes was kept constant. Electrodes were dried overnight at 100 °C under a vacuum before being transferred into an argon-filled glove box for cell assembly. The Li-O_2 test cell (EQ-STC-LI-AIR, MTI Corporation, Richmond, CA, USA) was constructed with lithium metal as the negative electrode and the MnO_2/C electrode as the positive electrode. A solution of 1 M $LiPF_6$ in a PC solvent was used as the electrolyte in all cells. After assembly, the test cell was taken away from the Ar-filled glove box and attached to a gas pipe that was constantly purged with dry O_2. Electrochemical tests were carried out after the cell was flushed with O_2 for 6 h. The cells were cycled galvanostatically with a BAT-750B (Acu Tech System, Taipei, Taiwan) at a constant current of 100 mA/g with a voltage region of 2.0–4.3 V *vs.* Li/Li^+ at room temperature.

For the RRDE experiments, the RRDE system (AFMT134DCPTT, Pine Research Instrumention, Durham, NC, USA) with interchangeable disk consisted of a 5 mm diameter glassy carbon electrode and a Pt ring electrode (1 mm width) with a 0.5 mm gap between them. The collection efficiency with this geometry is 0.24. The rotating ring-disk assembly was operated on a Pine AFMSRX rotator and CH705 Bipotentiostat (CH Instruments, Austin, TX, USA) with a computerized interface. Experiments were conducted using a three-electrode cell containing 10 mL of the electrolyte of interest and assembled in a dry Ar-filled AtmosBag (Sigma-Aldrich Z108450, St. Louis, MO, USA). Figure 1 shows the schematic of a four-neck, jacketed glass cell with the RRDE system. The counter electrode was a Li foil connected to a Ni wire, which was isolated by a layer of Celgard 2400 separator to prevent convective oxygen transport to the electrode. The Ag/Ag^+ reference electrode consisted of an Ag wire immersed into 0.1 M $AgNO_3$ in CH_3CN and sealed with a vycor frit at its tip. All potentials in this study were referenced to the Li/Li^+ potential scale (volts *vs.* Li^+/Li or V_{Li+}), obtained by calibration of the reference electrode against a fresh lithium wire before the experiments ($0\ V_{Li} = -3.46 \pm 0.01$ V *vs.* Ag/Ag^+). The working electrode consisted of a catalyst-covered glassy carbon disk and was immersed into the Ar or O_2-purged electrolyte for 30 min before each experiment. Prior to the RRDE measurements, Alternating current (AC) impedance measurements were carried out to determine the uncompensated ohmic electrolyte drop between working and reference electrodes by applying a 10 mV perturbation (0.1 MHz to 10 mHz) at the open circuit. IR (drop) correction to remove ohmic losses was performed by considering a total cell resistance of ~293 Ω measured by AC impedance. The capacitive-corrected ORR currents were calculated by subtracting the current measurement under Ar from that obtained in pure O_2 under identical scan rates, rotation speeds, and catalyst loadings.

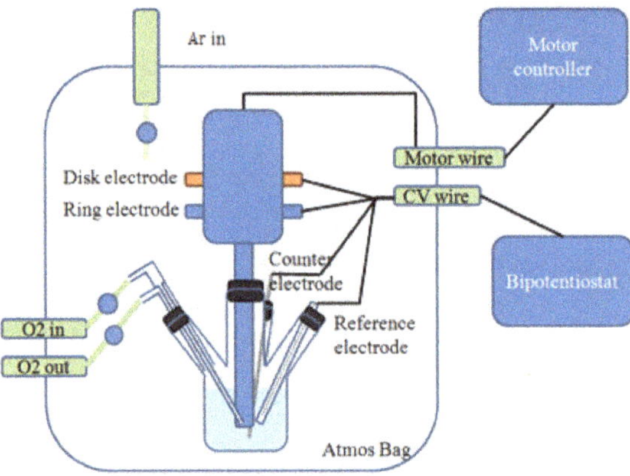

Figure 1. Schematic of a four-neck, jacketed glass cell with a rotating ring-disk electrode (RRDE) system.

3. Results and Discussion

The phase composition and structure of the prepared MnO_2/C composites were examined by the wide-angle XRD and SAXS patterns given in Figure 2a,b. As shown in Figure 2a, all peaks can be identified as a pure and well-crystallized β-MnO_2 phase (JCPDS 24-0735) with an ordered tetragonal structure indexed to the P42/mnm space group. Moreover, the XRD curves did not show any evidence of the formation of crystalline or amorphous carbon. It appears that when using resol/Pluronic F127 templates as a carbon source, the final product is most likely to remain amorphous or in a low crystalline carbon state. The appearance of the scattering peak in the SAXS pattern, as shown in Figure 2b, indicates the long-range regularity and highly ordered nature of the mesoporous structures of the prepared MnO_2/C composite.

The morphology of the prepared MnO_2/C composite was observed using SEM and TEM, as shown in Figure 3a–f. From the SEM images of the MnO_2/C composite (Figure 3a,b), it is clear that the oriented tetragonal MnO_2 nanorods are arranged on the surface of the carbon matrix. The prepared β-MnO_2 nanorods, typically 2–3 μm in length, have a square cross-section with an edge length in the range of 200–300 nm. Figure 3c,d show the TEM images of the MnO_2/C composite at different magnifications. Large domains of highly ordered stripe-like 1D channels are clearly observed. Figure 3e displays a TEM image of a typical nanorod with a smooth surface, and a SAED pattern based on a single nanorod (Figure 3f), indicating single-crystalline nature. The SEM and TEM analysis suggest that hollow MnO_2 nanorods grow homogeneously on the ordered mesoporous carbon frameworks to form the structure of the hierarchically mesoporous MnO_2/C composite.

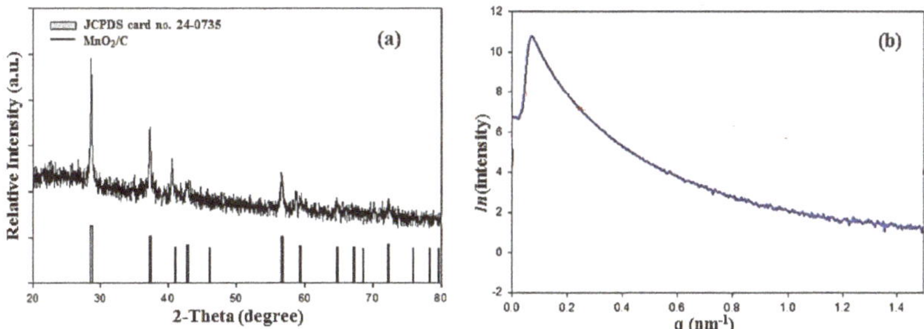

Figure 2. (**a**) Wide-angle X-ray diffraction (XRD) patterns; (**b**) small angle X-ray scattering (SAXS) patterns of MnO_2/C composites.

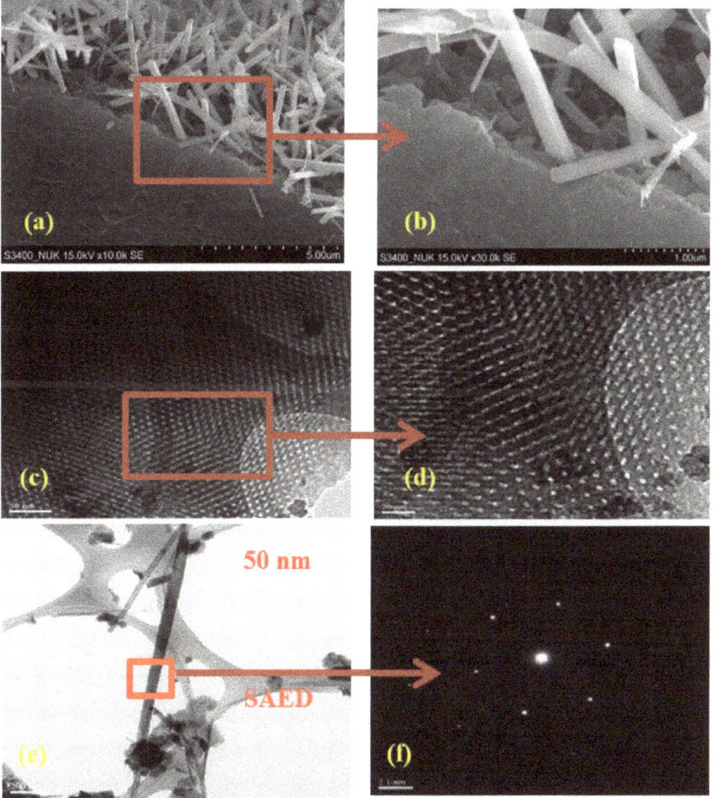

Figure 3. Scanning electron microscope (SEM) images (**a**) MnO_2/C composites; (**b**) high magnification of the region marked with a square in (**a**); and transmission electron microscope (TEM) images (**c**,**e**) MnO_2/C composites; (**d**) high magnification of the region marked with a square in (**c**); and (**f**) selected area electron diffraction (SAED) pattern of the region marked with a square in (**e**).

The pore structure of the mesoporous MnO_2/C composite was determined by nitrogen adsorption-desorption isothermal measurements. As shown in Figure 4, the adsorption isothermal curve of the MnO_2/C composite has a well-defined step as in typical IV classification with a H_1-type hysteretic loop in the p/p_o range of 0.40–1.0, indicating mesoporous material character. These findings suggest that the MnO_2/C composite sample does not contain framework-confined pores but is rather made up of individual nanorods. This is in agreement with the results from the SEM and TEM images. The Barrett–Joyner–Halenda (BJH) pore size distribution for the mesoporous MnO_2/C composite, shown in the insert of Figure 4, reveals peaks centering at 4.8 and 35 nm. This result confirms that most of the pore channels in the ordered mesoporous carbon are not blocked by the loading of MnO_2 nanorods. The nanoarchitecture of ordered mesoporous channels is maintained, which is desirable for the O_2 electrode in Li-O_2 batteries. Moreover, the measured BET surface area of the MnO_2/C composite is relatively high, at about 424 $m^2 \cdot g^{-1}$. The hierarchical microstructure of the MnO_2/C composite results in a large specific surface area. This is important for enhancing the electrochemical properties of an O_2 cathode material.

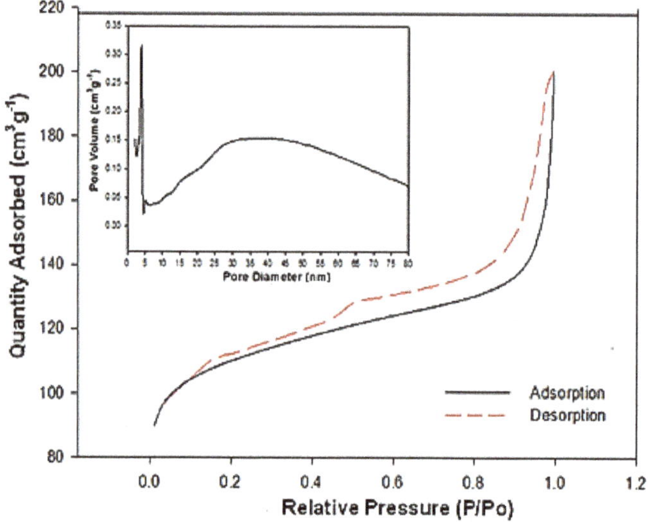

Figure 4. Nitrogen sorption isotherms of MnO_2/C composites. The insert is the Barrett–Joyner–Halenda (BJH) desorption pore size distribution.

MnO_2 has been known as a highly active ORR catalyst for some time [30,32,36] and has recently been applied as an O_2 cathode catalyst in Li-O_2 batteries. Due to the studies of the electrocatalytic activity of MnO_2, the following discussions regarding electrochemical tests make comparisons between Super-P carbon (SP) and MnO_2/C materials. To better study the catalytic activity of the electrodes, cyclic voltammetry (CV) and charge-discharge voltage measurements were carried out. At first, CV was carried out in the Ar-purged electrolyte and subsequently in the same solution saturated with O_2. The capacitively-corrected CV curves derived from both measurements are shown in Figure 5a. The CV plots of the O_2 electrodes prepared from MnO_2/C and SP cycled between 1.5 and 4.5 V with 2 $mV \cdot s^{-1}$ and the O_2-saturated 1 M $LiPF_6$/PC electrolyte are shown in Figure 5a. From the CV curves, the reduction peak voltage is shifted toward positive voltage, exhibiting electrocatalytic activity in the ORR of both samples. However, the MnO_2/C offers more positive onset reduction peak potential and a larger peak current, which clearly indicate the superior electrocatalytic activity of MnO_2/C compared to SP. Furthermore, the onset oxidation peaks appearing in the CV curves are about 2.7 and 2.9 V for MnO_2/C and SP, respectively. This demonstrates that MnO_2/C, with its lower onset oxidation peak, is more efficient for Li_2O_2 decomposition and has higher catalytic activity for the OER. The initial charge–discharge voltage profiles for both samples are shown in Figure 5b. The charge–discharge profiles of the MnO_2/C electrode exhibit much lower charge overpotential than do those of the SP electrode, although the reduction of the total overpotential is only about 25%. The round-trip efficiencies of the Li-O_2 batteries with a MnO_2/C electrode were lower than those with the SP electrode. These results indicate that the MnO_2/C composite can facilitate the complete reversibility of ORR and OER with low polarization for a Li-O_2 battery. This finding is in good agreement with the CV measurement. The initial discharge capacities of the MnO_2/C and SP electrodes were 612 mAh$\cdot$ g^{-1} and 589 mAh$\cdot$ g^{-1}, respectively. The good electrochemical performance of the MnO_2/C electrode may be due to the hierarchical mesostructure and large specific surface area, and the catalytic activity of the MnO_2/C composite.

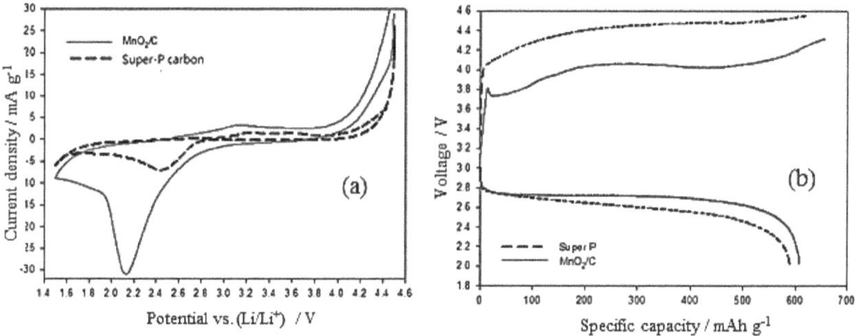

Figure 5. (**a**) CV curves were recorded at a scanning rate of 2 mV$\cdot$ s^{-1} for MnO_2/C and Super-P carbon samples; (**b**) initial charge–discharge profiles for MnO_2/C and Super P samples at a current density of 0.2 mA$\cdot$ cm^{-2}.

The rotating ring disk electrode (RRDE) technique was also used to investigate the kinetics of ORR since the ORR current is strongly relevant to hydrodynamic conditions [31]. Here, we used a glassy carbon (GC) electrode and an as-prepared MnO_2/C composite coated on the GC (MnO_2/C-GC) electrode as the working electrodes to study the stability of the electrolyte at the MnO_2/C electrode. Many reports [1,5,9] have shown the $O_2^{\bullet-}$ produced in the first step of the ORR upon battery discharge:

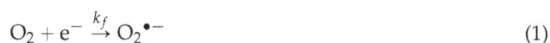

$$O_2 + e^- \xrightarrow{k_f} O_2^{\bullet-} \tag{1}$$

The reaction between the $O_2^{\bullet-}$ and the electrolyte is the critical problem that causes poor Li-O_2 battery cyclability. In the PC-based electrolyte, the ethereal carbon atom in PC suffers from nucleophilic attacks by $O_2^{\bullet-}$, yielding carbonate, acetate, and formate species (among others), according to Equation (2) [37,38]:

$$PC + O_2^{\bullet-} \xrightarrow{k} CO_3^{2-}, HCOO^-, CH_3COO^- \tag{2}$$

Here, we applied rotating disk electrode (RDE) voltammetry to measure the rate constant (k_f) when reducing O_2 to $O_2^{\bullet-}$ for Equation (1) in the 0.1 M $TBAPF_6$/PC electrolyte. The reaction rate constant, k_f, can be evaluated via the Koutecky–Levich (K–L) equation for a first-order reaction as follows [31]:

$$\frac{1}{i} = \frac{1}{i_k} + \frac{1}{i_d} \tag{3}$$

$$i_k = nFk_fC_{O_2} \tag{4}$$

$$i_d = 0.62nFD_{O_2}{}^{2/3}\nu^{-1/6}C_{O_2}\omega^{1/2} \tag{5}$$

where i_k and i_d represent kinetics and diffusion limiting current density ($A \cdot m^{-2}$), respectively; n is the number of electrons exchanged in the electrochemical reaction; F is Faraday's constant ($96{,}485\ C \cdot mol^{-1}$); k_f is the rate constant for Equation (1); D_{O_2} is the diffusion coefficient of O_2 in the solution; ν is the kinematic viscosity; ω is the angular frequency of the rotation; and C_{O_2} is the saturation concentration of O_2 in the solution. Additionally, knowing the values of ν and D_{O_2} for an electrolyte, one can obtain the concentration of oxygen (C_{O_2}) and rate constant (k_f) for Equation (1) by linearly fitting the K-L plots of i^{-1} *vs.* $\omega^{-0.5}$, as follows

$$\frac{1}{i} = \frac{1}{i_k} + \frac{1}{0.62nFD_{O_2}^{2/3}\nu^{-1/6}C_{O_2}\omega^{1/2}} \tag{6}$$

Prior to estimating the value of k_f from the K–L equation, the kinematic viscosity (ν) of the electrolyte and the diffusion coefficients of O_2 and $O_2^{\bullet-}$ (D_{O_2} and $D_{O_2^-}$), need to be quantified. The value of ν for PC with 0.1 M $TBAPF_6$ is $2.59 \times 10^{-2}\ cm^2 \cdot s^{-1}$ at 25 °C (ρ = 1.2 $g \cdot mL^{-1}$ and η = 3.13 mPa·s) and was measured by a Rheometer (Malvern Gemini, Malvern Instruments Ltd., Malvern, UK). For a known viscosity, the diffusion coefficients can be directly determined from the transit-time (T_s) measurement by the RRDE technique, as reported previously [37,39]. Figure 6a shows an example of T_s measurement in O_2-saturated solutions of 0.1 M $TBAPF_6$ in PC at ω = 100 rpm; T_s, the origin of which is taken at time = 2 s (the time at which the disk is conducted cathodic potential at 1.85 V_{Li}), is measured graphically from the intercept of the base steady ring current and the fast attenuate ring current line. Figure 6b,c show measurements of steady ring currents at the rotation rates (ω) of different electrodes, yielding T_s values for O_2 and $O_2^{\bullet-}$. Then, the obtained T_s is related to the ω and the ratio of ν and the diffusion coefficient (D), according to Equation (7) [37,39]:

$$T_s = K\left(\frac{\nu}{D}\right)^{1/3}\omega^{-1} \tag{7}$$

where K is proportionally constant depending on the RRDE's geometry; K = $43.1[\log(r_2/r_1)]^{2/3}$ (for T_s, reported in s and ω in rpm). For the RRDE used here, with r_1 = 0.25 cm and r_2 = 0.325 cm, the value of K is 10.1 rpm·s. Table 1 shows the estimated values of the diffusion coefficients of O_2 and $O_2^{\bullet-}$ calculated from Equation (7) based on the slopes of T_s *vs.* ω^{-1} obtained from Figure 6d. Figure 7a shows that well-defined O_2 diffusion-limited currents are obtained for the ORR on a GC electrode in an O_2-saturated 0.1 M $TBAPF_6$/PC solution. The K–L plot for the disk current values at 1.50 V_{Li} reveals the expected linear relation between the inverse of the limiting current and $\omega^{-0.5}$ (see Equation (6)). As shown in Table 1, the concentration of oxygen (C_{O_2}) on the GC electrode was estimated from the slope of the K–L plot using the prior measured values of ν and D_{O_2}, where n = 1 (according to the reaction of Equation (1)). The value of C_{O_2} is 6.1 M, which is higher than the finding of a previous report (4.8 M) [37]. This can be attributed to the larger O_2 flow rate in this experiment. The estimated value of C_{O_2} was also applied in the following calculations of the MnO_2/C-GC electrode since the same operation parameters (*i.e.*, O_2 flow rate, electrolyte composition, and amount) were used, as listed in Table 1. The rate constant for producing $O_2^{\bullet-}$, k_f for GC and the MnO_2/C-GC electrodes can be obtained by linearly fitting the K–L plots of i^{-1} *vs.* $\omega^{-0.5}$ (see Equation (6)), as shown in Figure 7a,b. The values of k_f for GC and the MnO_2/C-GC electrodes are $1.92 \times 10^{-2}\ cm \cdot s^{-1}$ and $4.29 \times 10^{-2}\ cm \cdot s^{-1}$, respectively. This result indicates that the MnO_2/C cathode catalyst exhibits a larger k_f value, resulting from higher electrocatalytic activity for the first step of the ORR (see Equation (1)) which produces a higher concentration of $O_2^{\bullet-}$.

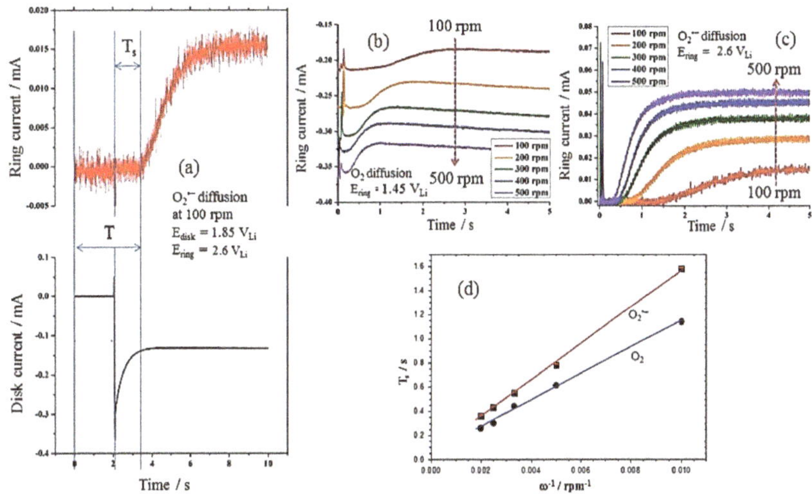

Figure 6. (**a**) Example of determination of the superoxide radical ($O_2^{\bullet-}$) transit-time (T_s) in O_2-saturated solutions of 0.1 M $TBAPF_6$ in propylene carbonate (PC) at ω = 100 rpm, E_{disk} = 1.85 V and E_{ring} = 2.6 V. Transit time (Ts) values at different rotation rates for the diffusion of (**b**) O_2 and (**c**) $O_2^{\bullet-}$; (**d**) relation between the inverse of the rotation speed and the transient time for O_2 and $O_2^{\bullet-}$.

Table 1. Summary of the electrolyte properties estimated with the proposed RRDE-based methodology and comparison with findings reported in the literature.

Disk Material/Electrolyte	ν ($cm^2 \cdot s^{-1}$)	D_{O_2} ($cm^2 \cdot s^{-1}$)	$D_{O_2^-}$ ($cm^2 \cdot s^{-1}$)	C_{O_2} (mM)	Reference
GC/0.1 M $TBAPF_6$, PC	2.6×10^{-2}	1.9×10^{-5}	8.6×10^{-6}	6.1	This work
MnO_2/C-GC/0.1 M $TBAPF_6$, PC	2.6×10^{-2}	1.9×10^{-5}	1.8×10^{-6}	6.1	This work
GC/0.2M TBATFSI, PC	2.6×10^{-2}	2.5×10^{-5}	6.8×10^{-6}	4.8	[37]

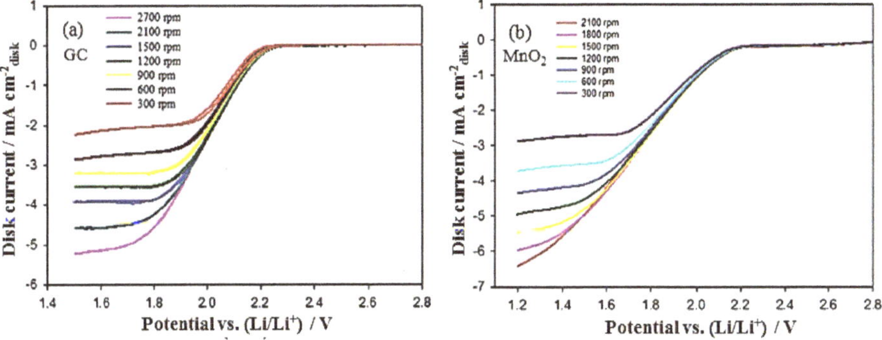

Figure 7. (**a**) Steady-state CV curves of a glassy carbon rotating disk electrode (RDE) in an O_2-saturated 0.1 M $TBAPF_6$/PC solution at a scan rate of 50 mV/s between 1.5 and 2.8 V_{Li} with different rotation rates. The insert is the Koutecky–Levich plot derived from the disc current values at 1.50 V_{Li}; (**b**) steady-state CV curves of a MnO_2/C RDE in an O_2-saturated 0.1 M $TBAPF_6$/PC solution at a scan rate of 50 mV/s between 1.2 and 2.8 V_{Li} with different rotation rates.

Recently, Herranz *et al.* [37] used RRDE voltammetry to quantify the stability of an electrolyte against $O_2^{\bullet-}$ by the rate constant (k) for Equation (2) According to their methods, the $O_2^{\bullet-}$ produced at the disk electrode in Equation (1) and the amount of $O_2^{\bullet-}$ were quantified at the ring electrode. The amount of $O_2^{\bullet-}$ consumed depends on the effective transient time, T_s, between the disk and the ring and the rate constant, k, for Equation (2). Longer T_s and larger k values cause increasing consumption of $O_2^{\bullet-}$ due to its reaction with the electrolyte, resulting in a lower $O_2^{\bullet-}$ oxidation current at the ring. Therefore, the collection efficiency, N_k, for $O_2^{\bullet-}$ at the ring electrode decreases with increasing transient time, which, in turn, depends on the geometry of ring and disk electrode, the diffusion coefficient of $O_2^{\bullet-}$ in the electrolyte, $D_{O_2^-}$, and the electrode rotation speed, ω. The correlation with the collection efficiency is the absolute ratio of ring and disk currents and can be characterized by the following equation [37,40]:

$$N_k = -\frac{i_{ring}}{i_{disk}} = N_{geometrical} - \beta^{\frac{2}{3}}\left(1 - UA_1^{-1}\right) + \frac{1}{2}A_1^{-1}A_2^2\kappa^2 U\beta^{\frac{4}{3}} - 2A_2\kappa^2 T_2 \quad (8)$$

where $A_1 = 1.288$, $A_2 = 0.643\ \nu^{1/6}\ D_{O_2^-}{}^{1/3}$, $\beta = 3\ln(r_3/r_2)$, $U = k^{-1}\tanh(A_1 k)$ and $T_2 = 0.718\ln(r_2/r_1)$, whereby r_1–r_3 refer to the radius of the disk and internal and external ring radii, respectively; ν is the kinematic viscosity; ω is the rotation rate; k is the rate constant for Equation (2); and $D_{O_2^-}$ is the diffusion coefficient of $O_2^{\bullet-}$. $N_{geometrical}$ is the geometrical collection efficiency of the RRDE corresponding to the fraction of a species electrochemically generated at the disk. This species is detected at the ring due to the lack of side-reactions with the electrolyte. Equation (8) shows the variation of N_k where the rotation rate and the rate constant (k) can be calculated at higher rotation rates, which show that the N_k is close to a constant value. Figure 8a shows the RRDE profiles of the MnO_2/C sample coating on the disk electrode. The disk and ring currents are recorded in an O_2-saturated 0.1 M $TBAPF_6$/PC solution at rotation rates between 300 and 2100 rpm, with continuous holding of the Pt ring at 2.85 V_{Li}. The ring current increases with the rotation rates because the shorter transient time at higher rotation rates reduces the reaction time between $O_2^{\bullet-}$ and the PC electrolyte so that a higher concentration of superoxide radical can be oxidized at the ring. Also, the N_k increases with rotation rates (ω) and is close to a constant value (0.14) at ω = 2100 rpm, as shown in Figure 8b. The PC-electrolyte decomposition rate constant (k) can be calculated by Equation (8) using the N_k value at a rotation speed of 2100 rpm with the kinematic viscosity (ν) and $D_{O_2^-}$ listed in Table 1. Table 2 shows the rate constant for producing $O_2^{\bullet-}$, k_f, and the PC-electrolyte decomposition rate constant, k, on the GC and MnO_2/C-GC electrodes. The value of k (1.5 s^{-1}) on the GC electrode is close to that of a previously reported measurement (k = 1.3 s^{-1}) [37]. Obviously, the k value on the MnO_2/C-GC electrode of 2.6 s^{-1} is larger than that on the GC electrode. This result shows that MnO_2/C is more active for the first step of the ORR (larger rate constant; k_f), producing a higher concentration of $O_2^{\bullet-}$ and leading to faster PC-electrolyte decomposition due to the attack by a large amount of $O_2^{\bullet-}$. Therefore, it is important to choose an appropriate electrolyte to avoid decomposition by $O_2^{\bullet-}$ attack for highly active catalyst applications on the cathode materials in Li-O_2 batteries. More detailed RRDE experiments and analysis will be carried out to estimate the decomposition rates of various electrolytes with different active catalysts.

Table 2. The rate constant for producing $O_2^{\bullet-}$, k_f, and the PC-electrolyte decomposition rate constant, k, on the GC and MnO_2/C-GC electrodes.

Disk Material/Electrolyte	k_f (cm·s^{-1})	k (s^{-1})	Reference
GC/0.1 M $TBAPF_6$, PC	1.9×10^{-2}	1.5	This work
MnO_2/C-GC/0.1 M $TBAPF_6$, PC	4.3×10^{-2}	2.6	This work
GC/0.2M TBATFSI, PC		1.3	[37]

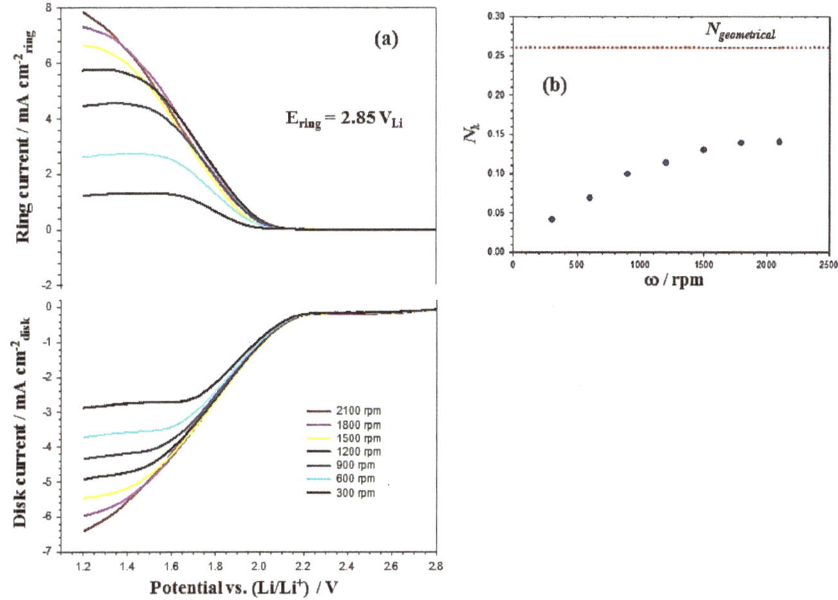

Figure 8. (**a**) RRDE profiles of MnO_2/C recorded at 50 mV·s^{-1} in an O_2-saturated 0.1 M $TBAPF_6$/PC solution, at rotation rates between 300 and 2100 rpm with continuous holding of the Pt ring at 2.85 V_{Li}; (**b**) evolution of the absolute ratio between the ring and disk current (N_k) and the electrode rotation speed (ω).

4. Conclusions

A hierarchically mesoporous carbon-supported manganese oxide (MnO_2/C) has been synthesized by a combination of soft template and hydrothermal methods. SEM and TEM analysis confirmed that hollow MnO_2 nanorods grow homogeneously on ordered mesoporous carbon frameworks to form a hierarchically mesoporous MnO_2/C composite structure. The CV and galvanostatic charge–discharge tests indicate that MnO_2/C composites have excellent catalytic activity towards ORR due to the larger surface area of ordered mesoporous carbon and higher catalytic activity of MnO_2.

The O_2 solubility, the diffusion rates of O_2 and $O_2^{\bullet-}$ coefficients, the rate constant for producing $O_2^{\bullet-}$ (k_f), and the PC-electrolyte decomposition rate constant (k) of the MnO_2/C composites have been measured by RRDE experiments and analysis in the 0.1 M $TBAPF_6$/PC electrolyte. The results indicate that MnO_2/C is more active for the first step of the ORR (larger rate constant; k_f), produces a higher concentration of $O_2^{\bullet-}$, and leads to faster PC-electrolyte decomposition due to the attack by a large amount of $O_2^{\bullet-}$. The stability of the electrolyte is very important when applying an active catalyst on a cathode material in Li-O_2 batteries. More detailed RRDE experiments and analysis will be carried out to estimate the decomposition rates of various electrolytes. These results seem to be interesting for the design of advanced Li-O_2 batteries with high electrochemical performance.

Acknowledgments: The authors thank the Ministry of Science and Technology of Taiwan for the financial support for this work under contract No. NSC-102-2113-M-390-005-MY3.

Author Contributions: Chih-Chun Chin and Hong-Kai Yang performed the experiments and analyzed the data. Jenn-Shing Chen contributed to the idea, discussion and writing the manuscript.

Conflicts of Interest: The authors declare no conflict of interest.

References

1. Bhatt, M.D.; Geaney, H.; Nolan, M.; O'Dwyer, C. Key scientific challenges in current rechargeable non-aqueous Li-O_2 batteries: Experiment and theory. *Phys. Chem. Chem. Phys.* **2014**, *16*, 12093–12130. [CrossRef] [PubMed]
2. Rezvanizaniani, S.M.; Liu, Z.; Chen, Y.; Lee, J. Review and recent advances in battery health monitoring and prognostics technologies for electric vehicle (EV) safety and mobility. *J. Power Sources* **2014**, *256*, 110–124. [CrossRef]
3. Lu, J.; Li, L.; Park, J.B.; Sun, Y.K.; Wu, F.; Amine, K. Aprotic and aqueous Li-O_2 batteries. *Chem. Rev.* **2014**, *114*, 5611–5640. [CrossRef] [PubMed]
4. Etacheri, V.; Sharon, D.; Garsuch, A.; Afri, M.; Frimer, A.A.; Aurbach, D. Hierarchical activated carbon microfiber (ACM) electrodes for rechargeable Li-O_2 batteries. *J. Mater. Chem. A* **2013**, *1*, 5021–5030. [CrossRef]
5. Capsoni, D.; Bini, M.; Ferrari, S.; Quartarone, E.; Mustarelli, P. Recent advances in the development of Li-air batteries. *J. Power Sources* **2012**, *220*, 253–263. [CrossRef]
6. Ding, N.; Chien, S.W.; Hor, T.S.A.; Lum, R.; Zong, Y.; Liu, Z. Influence of carbon pore size on the discharge capacity of Li-O_2 batteries. *J. Mater. Chem. A* **2014**, *2*, 12433–12441. [CrossRef]
7. Bruce, P.G.; Freunberger, S.A.; Hardwick, L.J.; Tarascon, J.M. Li-O_2 and Li-S batteries with high energy storage. *Nat. Mater.* **2012**, *11*, 19–29. [CrossRef] [PubMed]
8. Yoon, T.H.; Park, Y.J. New strategy toward enhanced air electrode for Li-air batteries: apply a polydopamine coating and dissolved catalyst. *RSC Adv.* **2014**, *4*, 17434–17442. [CrossRef]
9. Xu, L.; Ma, J.; Li, B.; Kang, F. A novel air electrode design: A key to high rate capability and long life span. *J. Power Sources* **2014**, *255*, 187–196. [CrossRef]
10. Wang, J.; Li, Y.; Sun, X. Challenges and opportunities of nanostructured materials for aprotic rechargeable lithium-air batteries. *Nano Energy* **2013**, *2*, 443–467. [CrossRef]
11. Padbury, R.; Zhang, X. Lithium-oxygen batteries-limiting factors that affect performance. *J. Power Sources* **2011**, *196*, 4436–4444. [CrossRef]
12. Lu, Y.C.; Gasteiger, H.A.; Shao-Horn, Y. Catalytic activity trends of oxygen reduction reaction for nonaqueous Li-air batteries. *J. Am. Chem. Soc.* **2011**, *133*, 19048–19051. [CrossRef] [PubMed]
13. Park, H.W.; Lee, D.U.; Nazar, L.F.; Chen, Z. Oxygen reduction reaction using MnO_2 nanotubes/nitrogen-doped exfoliated graphene hybrid catalyst for Li-O_2 battery applications. *J. Electrochem. Soc.* **2013**, *160*, A344–A350. [CrossRef]
14. Zhang, Z.; Zhou, G.; Chen, W.; Lai, Y.; Li, J. Facile synthesis of Fe_2O_3 nanoflakes and their electrochemical properties for Li-air batteries. *ECS Electrochem. Lett.* **2014**, *3*, A8–A10. [CrossRef]
15. Li, J.; Zhao, Y.; Zou, M.; Wu, C.; Huang, Z.; Guan, L. An Effective Integrated Design for Enhanced Cathodes of Ni Foam-Supported Pt/Carbon Nanotubes for Li-O_2 Batteries. *ACS Appl. Mater. Interfaces* **2014**, *6*, 12479–12485. [CrossRef] [PubMed]
16. McCloskey, B.D.; Scheffler, R.; Spidel, A.; Girishkumar, G.; Luntz, A.C. On the mechanism of nonaqueous Li-O_2 electrochemistry on C and its kinetic overpotentials: some implications for Li-air batteries. *J. Phys. Chem. C* **2012**, *116*, 23897–23905. [CrossRef]
17. Meini, S.; Solchenbach, S.; Piana, M.; Gasteiger, H.A. The role of electrolyte solvent stability and electrolyte impurities in the electrooxidation of Li_2O_2 in Li-O_2 batteries. *J. Electrochem. Soc.* **2014**, *161*, A1306–A1314. [CrossRef]
18. Lu, Y.-C.; Gasteiger, H.A.; Shao-Horn, Y. Method development to evaluate the oxygen reduction activity of high-surface-area catalysts for Li-air batteries. *Electrochem. Solid-State Lett.* **2011**, *14*, A70–A74. [CrossRef]
19. Sun, B.; Huang, X.; Chen, S.; Munroe, P.; Wang, G. Porous graphene nanoarchitectures: An efficient catalyst for low charge-overpotential, long life, and high capacity lithium-oxygen batteries. *Nano Lett.* **2014**, *14*, 3145–3152. [CrossRef] [PubMed]
20. Chervin, C.N.; Wattendorf, M.J.; Long, J.W.; Kucko, N.W.; Rolison, D.R. Carbon nanofoam-based cathodes for Li-O_2 batteries: Correlation of pore-solid architecture and electrochemical performance. *J. Electrochem. Soc.* **2013**, *160*, A1510–A1516. [CrossRef]
21. Lu, F.; Cao, X.; Wang, Y.; Jin, C.; Shen, M.; Yang, R. A hierarchical $NiCo_2O_4$ spinel nanowire array as an electrocatalyst for rechargeable Li-air batteries. *RSC Adv.* **2014**, *4*, 40373–40376. [CrossRef]

22. Xiao, J.; Mei, D.; Li, X.; Xu, W.; Wang, D.; Graff, G.L.; Bennett, W.D.; Nie, Z.; Saraf, L.V.; Aksay, I.A.; *et al.* Hierarchically porous graphene as a lithium-air battery electrode. *Nano Lett.* **2011**, *11*, 5071–5078. [CrossRef] [PubMed]
23. Cao, R.; Thapa, R.; Kim, H.; Xu, X.; Kim, M.G.; Li, Q.; Park, N.; Liu, M.; Cho, J. Promotion of oxygen reduction by a bio-inspired tethered iron phthalocyanine carbon nanotube-based catalyst. *Nature Commun.* **2013**, *4*, 2076–2082. [CrossRef] [PubMed]
24. Guo, Z.; Zhou, D.; Dong, X.; Qiu, Z.; Wang, Y.; Xia, Y. Ordered hierarchical mesoporous/macroporous carbon: A high-performance catalyst for rechargeable Li-O_2 batteries. *Adv. Mater.* **2013**, *25*, 5668–5672. [CrossRef] [PubMed]
25. Sun, B.; Liu, H.; Munroe, P.; Ahn, H.; Wang, G. Nanocomposites of CoO and a mesoporous carbon (CMK-3) as a high performance cathode catalyst for lithium-oxygen batteries. *Nano Res.* **2012**, *5*, 460–469. [CrossRef]
26. Wang, Z.-L.; Xu, D.; Xu, J.-J.; Zhang, L.-L.; Zhang, X.-B. Graphene oxide gel-derived, free-standing, hierarchically porous carbon for high-capacity and high-rate rechargeable Li-O_2 batteries. *Adv. Funct. Mater.* **2012**, *22*, 3699–3705. [CrossRef]
27. Li, Y.; Wang, J.; Li, X.; Geng, D.; Banis, M.N.; Li, R.; Sun, X. Nitrogen-doped graphene nanosheets as cathode materials with excellent electrocatalytic activity for high capacity lithium-oxygen batteries. *Electrochem. Commun.* **2012**, *18*, 12–15. [CrossRef]
28. Cheng, H.; Scott, K. Carbon-supported manganese oxide nanocatalysts for rechargeable lithium–air batteries. *J. Power Sources* **2010**, *195*, 1370–1374. [CrossRef]
29. Park, H.W.; Lee, D.U.; Liu, Y.; Wu, J.; Nazar, L.F.; Chen, Z. Bi-functional N-doped CNT/graphene composite as highly active and durable electrocatalyst for metal air battery applications. *J. Electrochem. Soc.* **2013**, *160*, A2244–A2250. [CrossRef]
30. Yu, Y.; Zhang, B.; Xu, Z.-L.; He, Y.-B.; Kim, J.-K. Free-standing Ni mesh with *in-situ* grown MnO_2 nanoparticles as cathode for Li-air batteries. *Solid State Ionics* **2014**, *262*, 197–201. [CrossRef]
31. Xiao, W.; Wang, D.; Lou, X.W. Shape-controlled synthesis of MnO_2 nanostructures with enhanced electrocatalytic activity for oxygen reduction. *J. Phys. Chem. C* **2010**, *114*, 1694–1700. [CrossRef]
32. Debart, A.; Paterson, A.J.; Bao, J.; Bruce, P.G. α-MnO_2 nanowires: A catalyst for the O_2 electrode in rechargeable lithium batteries. *Angew. Chem.* **2008**, *47*, 4521–4524. [CrossRef] [PubMed]
33. Landa-Medrano, I.; Pinedo, R.; de Larramendi, I.R.; Ortiz-Vitoriano, N.; Rojo, T. Monitoring the location of cathode-reactions in Li-O_2 batteries. *J. Electrochem. Soc.* **2015**, *162*, A3126–A3132. [CrossRef]
34. Huang, C.-H.; Gu, D.; Zhao, D.; Doong, R.-A. Direct synthesis of controllable microstructures of thermally stable and ordered mesoporous crystalline titanium oxides and carbide/carbon composites. *Chem. Mater.* **2010**, *22*, 1760–1767. [CrossRef]
35. Meng, Y.; Gu, D.; Zhang, F.; Shi, Y.; Yang, H.; Li, Z.; Yu, C.; Tu, B.; Zhao, D. Ordered mesoporous polymers and homologous carbon frameworks: Amphiphilic surfactant templating and direct transformation. *Angew. Chem.* **2005**, *44*, 7053–7059. [CrossRef] [PubMed]
36. Wang, Z.-L.; Xu, D.; Xu, J.-J.; Zhang, X.-B. Oxygen electrocatalysts in metal-air batteries: From aqueous to nonaqueous electrolytes. *Chem. Soc. Rev.* **2014**, *43*, 7746–7786. [CrossRef] [PubMed]
37. Herranz, J.; Garsuch, A.; Gasteiger, H.A. Using rotating ring disc electrode voltammetry to quantify the superoxide radical stability of aprotic Li-air battery electrolytes. *J. Phys. Chem. C* **2012**, *116*, 19084–19094. [CrossRef]
38. Freunberger, S.A.; Chen, Y.; Peng, Z.; Griffin, J.M.; Hardwick, L.J.; Barde, F.; Novak, P.; Bruce, P.G. Reactions in the rechargeable lithium-O_2 battery with alkyl carbonate electrolytes. *J. Am. Chem. Soc.* **2011**, *133*, 8040–8047. [CrossRef] [PubMed]
39. Chatenet, M.; Molina-Concha, M.B.; El-Kissi, N.; Parrour, G.; Diard, J.P. Direct rotating ring-disk measurement of the sodium borohydride diffusion coefficient in sodium hydroxide solutions. *Electrochim. Acta* **2009**, *54*, 4426–4435. [CrossRef]
40. Bard, A.J.; Faulkner, L.R. *Electrochemical Methods: Fundamentals and Applications*; John Wiely & Sons: New York, NY, USA, 2001; p. 519.

Article

The Influence of Carbonaceous Matrices and Electrocatalytic MnO_2 Nanopowders on Lithium-Air Battery Performances

Alessandro Minguzzi [1,2], Gianluca Longoni [3], Giuseppe Cappelletti [1,4,*], Eleonora Pargoletti [1,4], Chiara Di Bari [5], Cristina Locatelli [1], Marcello Marelli [6], Sandra Rondinini [1,2,4] and Alberto Vertova [1,2,4]

1 Dipartimento di Chimica, Università degli Studi di Milano, via Golgi 19, 20133 Milano, Italy; alessandro.minguzzi@unimi.it (A.M.); eleonora.pargoletti@gmail.com (E.P.); cristina.locatelli@unimi.it (C.L.); sandra.rondinini@unimi.it (S.R.); alberto.vertova@unimi.it (A.V.)

2 ISTM-CNR, Istituto di Scienze e Tecnologie Molecolari, c/o Dipartimento di Chimica, Università degli Studi di Milano, via Golgi 19, 20133 Milano, Italy

3 Dipartimento di Scienza dei Materiali, Università degli Studi di Milano Bicocca, via Roberto Cozzi 55, 20125 Milano, Italy; gianluca.longoni@mater.unimib.it

4 Consorzio Interuniversitario Nazionale per la Scienza e Tecnologia dei Materiali—INSTM, via G. Giusti 9, 50121 Firenze, Italy

5 Istituto de Catalisis y Petroleoquimica, Consejo Superior de Investigaciones Cientificas, C/Marie Curie 2, L10, 28049 Madrid, Spain; cdi.bari@csic.es

6 CNR-ISTM/ISTeM, via Fantoli 15/16, 20138 Milano, Italy; m.marelli@istm.cnr.it

* Correspondence: giuseppe.cappelletti@unimi.it; Tel./Fax: +39-02-5031-4228

Academic Editors: Hermenegildo García and Sergio Navalón
Received: 1 December 2015; Accepted: 31 December 2015; Published: 6 January 2016

Abstract: Here, we report new gas diffusion electrodes (GDEs) prepared by mixing two different pore size carbonaceous matrices and pure and silver-doped manganese dioxide nanopowders, used as electrode supports and electrocatalytic materials, respectively. MnO_2 nanoparticles are finely characterized in terms of structural (X-ray powder diffraction (XRPD), energy dispersive X-ray (EDX)), morphological (SEM, high-angle annular dark field (HAADF)-scanning transmission electron microscopy (STEM)/TEM), surface (Brunauer Emmet Teller (BET)-Barrett Joyner Halenda (BJH) method) and electrochemical properties. Two mesoporous carbons, showing diverse surface areas and pore volume distributions, have been employed. The GDE performances are evaluated by chronopotentiometric measurements to highlight the effects induced by the adopted materials. The best combination, hollow core mesoporous shell carbon (HCMSC) with 1.0% Ag-doped hydrothermal MnO_2 (M_hydro_1.0%Ag) allows reaching very high specific capacity close to 1400 $mAh \cdot g^{-1}$. Considerably high charge retention through cycles is also observed, due to the presence of silver as a dopant for the electrocatalytic MnO_2 nanoparticles.

Keywords: manganese dioxide nanoparticles; silver doping; mesoporous carbon; gas diffusion electrode (GDE); Li-air battery

---

1. Introduction

Electrochemical power suppliers, capable of storing high quantities of energy, have always been one of the most challenging topics in electrochemistry. The improvement of their efficiency, especially in terms of their energy density, could lead to the exploitation of these devices in the automotive industry and in the correct development of renewable energy sources, giving the possibility to overcome the discontinuous energy production. Recently, thanks to the exploratory work by Abraham *et al.* [1], a new device (Li-O_2 battery) able to give high theoretical energy densities has begun to be investigated. It is a secondary battery, akin to the familiar metal-air devices, like zinc-air batteries: (($-$) Li/non-aqueous electrolyte/O_2 (air) (+)). It is noteworthy to remember that the oxidation of 1 kg of lithium releases around 12 kWh, a value comparable to the theoretical energy density of gasoline [2].

One of the most challenging features in the cathodic compartment of metal-air batteries is the composite gas diffusion electrode (GDE), which has to be permeable by oxygen in alternating directions (discharge/charge reactions). Thus, air-cathode structures and the employed material morphologies are crucial to assess from both electrochemical [3–5] and morphological/structural [6] techniques. Moreover, in these systems, current density, electrolyte composition and discharge products (Li_2O_2 (E = 3.10 V *vs.* Li^+/Li) and the subsequent reduced species Li_2O (E = 2.72 V *vs.* Li^+/Li) [7,8]) are important parameters that can seriously affect the performance of the entire battery [9–11]. In particular, the air-cathode (the oxygen reduction site) requires a complex combination of electrocatalytic materials, supports and a current collector in order to optimize the formation of an extended triple contact and the fast and reversible formation/removal of the cathode reaction products. Recent literature studies have demonstrated how the specific capacity has a stronger reliance on the pore size distribution of the cathode material (typically graphitic materials) than a direct correlation with the total surface areas [12–14]. Meso- and macro-porosity allows insoluble LiO_2 and Li_2O_2 discharge products to homogeneously fill the pores, whereas micro-pores become quickly top-clogged by solid particles, determining the fast decay in the active surface area of the material [15,16]. Aiming at obtaining an optimal response between discharge and charge curves, suitable electrocatalysts have to be included into the cathode structure, in order to reduce and equalize the overpotentials for the oxygen reduction/oxidation. Thus, many transition metals and their relative oxides, such as Au, Pt, NiO, Fe_2O_3 and Fe_3O_4 in aprotic media [17] and IrO_2-SnO_2 mixtures in alkaline protic solvent [18,19], have been investigated. Moreover, $CoFe_2O_4$ and CuO nanoparticles have been also tested giving the best capacity retention properties [17]. In this context, new approaches have been developed exploiting a solution-phase catalyst in order to catalyze the Li_2O_2 decomposition during the charge cycle [20–22]. The most promising material, in terms of performances in both oxygen reduction (discharge) and evolution (charge) and costs, seems to be manganese dioxide nanoparticles. According to the literature, MnO_2 would ensure capacities comparable to those of platinum, letting higher capacity retention to be reached [23,24], even in the presence of non-aqueous electrolytes, widely used to prevent Li decomposition. However, these non-aqueous electrolytes can be affected by electrode surface potentials, causing a rapid degradation of the electrolyte itself and leading to other discharge products (lithium alkyl carbonates or simply Li_2CO_3) [25–27]. The usage of propylene carbonate (PC), and in general of organic carbonates, is still an open debate and studies on the mechanism and by-product formation of carbonate solvent degradation in Li/air batteries are going on [27]. To overcome this problem, aprotic electrolytes (such as DMSO or tetraethylene glycol dimethyl ether (TEGDME)) have been used lately in Li-O_2 batteries [28–30]. However, also these newly-adopted solvents show some drawbacks, *i.e.*, unwanted reactions leading to the consequent formation of by-products, such as, for example, Li_2CO_3 and LiOH [31,32].

In the present research work, the electrocatalytic activity of different hydrothermally-synthesized MnO_2, supported on *ad hoc* home-made mesoporous carbons, is evaluated using two different lithium-air cell configurations. Correlations between the physico-chemical characteristics of the materials, employed to prepare GDEs and the final electrical performances of the cell, are drawn. Moreover, taking into account all of the shortcomings related to the use of non-aqueous electrolytes [17,33,34], $LiClO_4$ in PC (a low cost material) has been employed aiming at evaluating the performance of both pure and doped manganese dioxide-based nanomaterials, as electrocatalysts.

2. Results and Discussion

2.1. Morphological and Structural Characterization of MnO_2 Nanomaterials

MnO_2 powders are widely used as cathodic material in batteries [16,17,35]; their electrochemical reactivity generally depends on morphological (surface area, size and type of pores, particle size) and structural (crystalline phases, presence of defects (microtwinning and De Wolff disorder) [36,37]) properties. Among the possible MnO_2 polymorphs, the γ-form, which consists of an intergrowth structure of β-pyrolusite and β-ramsdellite [35,37], generally exhibits the highest electrochemical reactivity [36–38]. The physico-chemical properties of γ-MnO_2 vary considerably with synthetic procedures and experimental conditions [39–41]. Herein, we adopted a hydrothermal route to prepare pure and Ag-doped samples.

The SEM image of hydrothermal MnO_2 (M_hydro) (Figure S1a) shows the micrometric spherical aggregates (2–5 μm) of MnO_2 composed by nanosized tiny sticks, with diameters in the range of 20–50 nm and lengths up to several hundred nanometers (Figure 1a,b). This result is fully in agreement with what was reported by Benhaddad *et al.* [38], who synthesized powders consisting of assembled straight needles characterized by similar average sizes. A further thermal treatment leads to the formation of bigger aggregates (up to 10 μm), as can be seen in Figure S1b in the case of particles calcined at 500 °C. After the thermal treatment, the elongated shapes are retained (Figure 1c,d), but their morphology strongly changes: several structural defects appear and the needles, characterized by greater sizes, become smoother with respect to the bare ones.

The surface area, the total pore volume and the relative diameter of mesopores between 6 and 20 nm are reported in Table 1 for all of the synthesized nanomaterials.

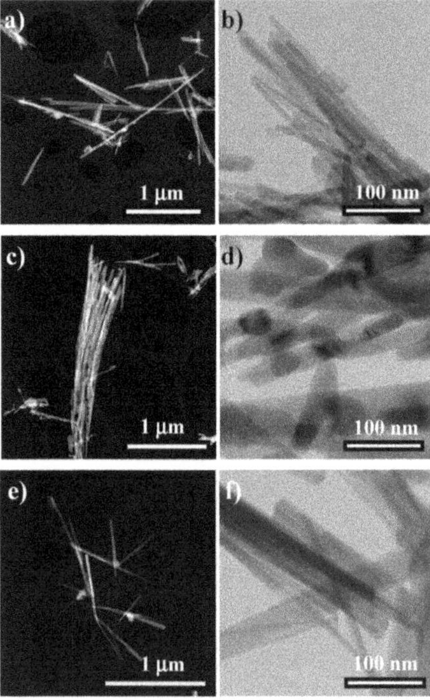

Figure 1. High-angle annular dark field (HAADF)-scanning transmission electron microscopy (STEM) (on the left) and transmission electron microscope (TEM) (on the right) images of (**a**,**b**) hydrothermal MnO_2 (M_hydro), (**c**,**d**) M_500 (500 °C) and (**e**,**f**) M_hydro_1.0%Ag samples.

Table 1. Specific surface area (S_{BET}), total pore volume ($V_{tot.pores}$) and relative percentage of the pore size with a diameter (d) between 6 and 20 nm for hydrothermal, calcined and doped MnO_2 samples.

Sample	S_{BET} ($m^2 \cdot g^{-1}$)	$V_{tot.pores}$ ($mL \cdot g^{-1}$)	6 nm < d < 20 nm
M_hydro	97	0.336	49
M_200	70	0.391	40
M_300	61	0.360	39
M_400	46	0.310	19
M_500	28	0.290	2
M_hydro_0.5%Ag	88	0.325	51
M_hydro_1.0%Ag	75	0.300	50
M_hydro_2.0%Ag	73	0.248	49

The hydrothermal powder (M_hydro) shows the highest surface area (97 $m^2 \cdot g^{-1}$) and the largest amount of pores with d < 20 nm, a typical porosity among the nanoneedles making the sticks [38]. The calcination temperature provokes a drastic decrease in the surface area of MnO_2 powders, which is due to the progressive sintering of particles and the collapse of pores, especially for the smallest ones (see Table 1, fourth column). According to these results, it can be noted that the heating step has a great effect on the pore size distribution of the synthesized MnO_2 powders. In order to elucidate this effect, Figure 2 shows the pore size distribution for the present samples determined by the Barrett Joyner Halenda (BJH) method. The average pore size shifts from 10 nm for the hydrothermal sample to a higher value (around 20–30 nm) for the samples calcined in the range of 200–400 °C. At 500 °C, a second population appears (~80 nm) corresponding to pores between sticks observed inside the MnO_2 bowls, as already reported in the literature [38].

From the structural point of view, many polymorphs are present in the M_hydro powder. Notwithstanding the absence of a heating procedure, the sample is well crystallized due to Ostwald ripening mechanisms of dissolution/precipitation typical of hydrothermal growth. Figure 3 shows this complex scenario, characterized by the presence of several crystalline phases. All of the reflections (with the *hkl* values and the principal relative intensity) of the X-ray diffraction line can be mainly indexed to four crystallographic polymorphs: (i) γ-MnO_2 nsutite (International Centre for Diffraction Data Powder Diffraction File (ICDD PDF-2) Card No. 14-0644) [36]; (ii) β-MnO_2 ramsdellite (ICDD PDF-2 Card No. 04-0378) [42]; (iii) β-MnO_2 pyrolusite (ICDD PDF-2 Card No. 04-0591) [43]; and (iv) α-MnO_2 hollandite (ICDD PDF-2 Card No. 44-0141) [44].

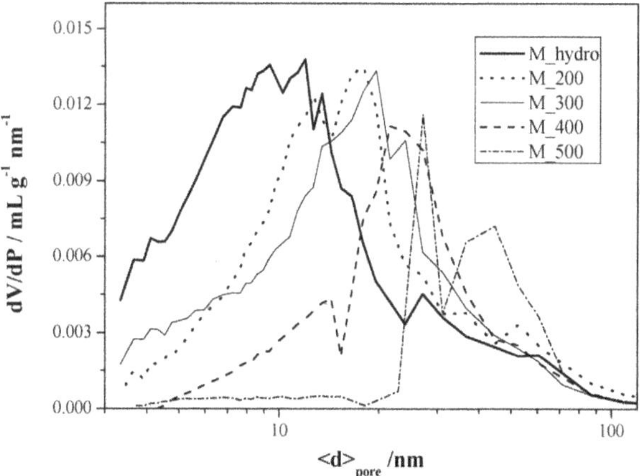

Figure 2. Comparison of the pore size distribution (calculated by the Barrett Joyner Halenda (BJH) method) of MnO_2 powders at increasing calcination temperature.

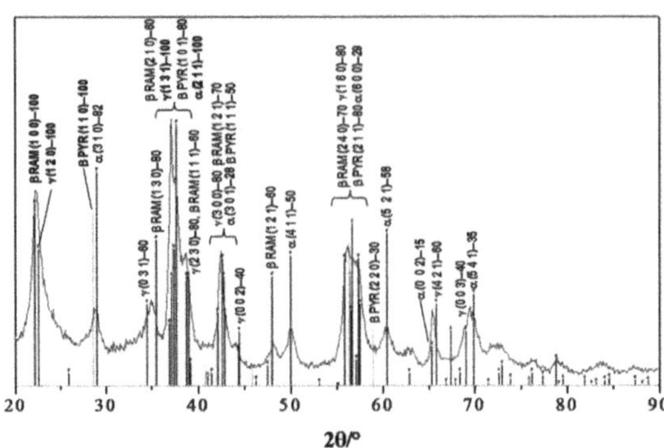

Figure 3. X-ray powder diffraction (XRPD) pattern of the M_hydro sample with the most intense reflections (*hkl*, intensity) of the main polymorphs (γ-MnO_2 nsutite, β-MnO_2 ramsdellite (RAM), β-MnO_2 pyrolusite (PYR) and α-MnO_2 hollandite).

Figure 4 shows that high calcination temperatures ($T > 300$ °C) lead to an increase of β-pyrolusite (100% reflection at around 28°, characterized by a low fraction of defects [38]) content to the detriment of the more electrochemically-active γ-MnO_2 form. At around 28°, also the α-MnO_2 polymorph could be present (the 310 plane, see Figure 3), but at higher temperatures, small traces are appreciable (see the peak at 50° ascribable only to the 411 reflection of α-MnO_2 phase). Finally, at 500 °C, a new phase appears indexed as bixbite Mn_2O_3 (ICDD PDF-2 Card No. 71-0636) [45]. On the basis of both structural and morphological results, it can be affirmed that the heating temperature engenders the transformation of γ-MnO_2 to β-pyrolusite and provokes the decrease of the surface area and pore volume, leading to higher crystallinity of the samples. Thus, for the evaluation of the performances of the final Li/Air battery, only M_hydro was used for the fabrication of the GDE cathodes. All of the calcined samples show very poor electrocatalytic activity: particularly, GDE-H-M_200, formed by hollow core mesoporous shell carbon (HCMSC) and MnO_2 calcined at 200 °C, gives a specific capacity less than 1000 mAh·g^{-1} at the first discharge run, which rapidly decreases, within four cycles, to a value close to zero, losing about 99% of its specific capacity after the first cycle.

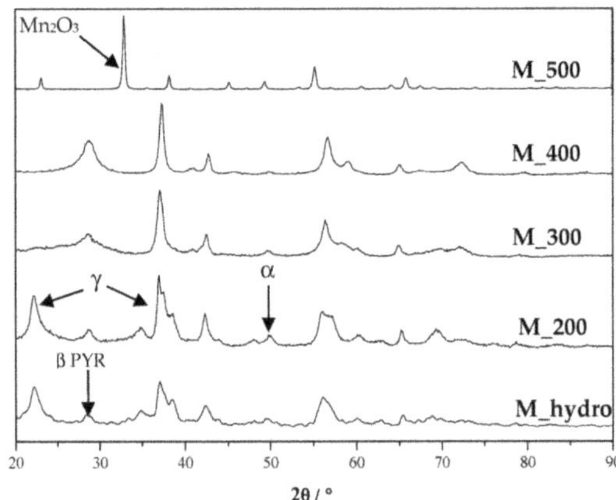

Figure 4. XRPD patterns of MnO_2 samples calcined at different temperatures. The most significant reflections for γ-MnO_2 nsutite, β-MnO_2 pyrolusite, α-MnO_2 hollandite and Mn_2O_3 bixbite are highlighted.

Regarding the hydrothermal samples doped with silver ions, three different molar ratio percentages (0.5%, 1.0%, 2.0%) were adopted. All of the samples show lower surface area and pore volume with respect to the undoped ones, but the same pore size distribution (Table 1, fourth column). Furthermore, the addition of Ag leads to phase composition, needles (Figure 1e,f) and an aggregate size (Figure S1c) comparable to the hydrothermal sample. From structural point of view, the increase of Ag content does not modify both the crystallinity and the lattice parameters, as shown in Figure S2, for bare and 1.0% Ag-doped samples (the latter as a representative sample of differently-doped powders). These effects may be due to a mild modification in the MnO_2 lattice parameters, induced by Ag atoms. Among the dopant concentrations, EDX mapping (Figure S1d) shows that the 1.0% molar ratio (M_hydro_1.0%Ag) can be considered the best homogeneous dopant distribution in the MnO_2 matrix; for this reason, only the latter powder was used in the lithium-air cell prototype.

2.2. Electrochemical Characterization of GDEs

GDEs, listed in Table 2 and prepared using mesoporous carbons (MCC, mesocellular carbon or HCMSC, hollow core mesoporous shell carbon, as supports) and MnO_2 powders (as the electrocatalysts), were investigated as described in the experimental part. They were tested at least three times in the same conditions in order to highlight both the reproducibility of the preparation route and the effects of the macroscopic morphology and of the surface texture on the electrochemical performances.

Table 2. Investigated gas diffusion electrodes (GDEs) with the relevant chemical compositions and active material loading. H: HCMSC (hollow core mesoporous shell carbon); M: MCC (mesocellular carbon); V: Vulcan XC72R; PVDF: polyvinylidene fluoride; SAB: Shawinigan Black AB50 carbon.

Name	Composition	Loading ($mg \cdot cm^{-2}$)
GDE-V	PVDF (15%), SAB (20%), Vulcan XC72R (65%)	1.3
GDE-V-Pt	PVDF (15%), SAB (20%), 10% Pt-loaded Vulcan XC72R (65%)	1.3
GDE-M	PVDF (15%), SAB (20%), MCC (65%)	1.1
GDE-M-M_hydro	PVDF (15%), SAB (20%), MCC (45%), M_hydro (20%)	2.2
GDE-H	PVDF (15%), SAB (20%), HCMSC (65%)	1.4
GDE-H-M_hydro	PVDF (15%), SAB (20%), HCMSC (45%), M_hydro (20%)	1.3
GDE-H-M_hydro_1.0%Ag	PVDF (15%), SAB (20%), HCMSC (45%), M_hydro_1.0%Ag (20%)	1.3

Figure 5a compares GDEs composed of the different synthesized mesoporous carbon supports and the reference one (GDE-V, Curve 1) during the first discharge cycle, using the S-cell configuration (Figure S3a). The reference support, GDE-V (Figure 5a, Curve 1), shows an inhomogeneous discharge profile with a plateau at 2.45 V and a very low specific capacity of about 900 $mAh \cdot g^{-1}$, consistent with the literature data [46]. All of the specific capacities are calculated by dividing the charge amount by active material weight, that is the sum of carbonaceous support and electrocatalytic nanopowders, when present. GDE-H (Figure 5a, Curve 2) and GDE-M (Figure 5a, Curve 3), exploiting HCMSC and MCC carbon matrices, respectively, behave better than either the reference cathodes or the newly-proposed material, like graphene [47,48]. MCC allows reaching a specific capacity of 1560 $mAh \cdot g^{-1}$, while the HCMSC-based cathode gives 1150 $mAh \cdot g^{-1}$. This different behavior is related to the pore distribution characteristics of the material: finer pores, less than 40 Å wide (HCMSC), hinder an easy growth and accommodation of the discharge products (Li-oxides).

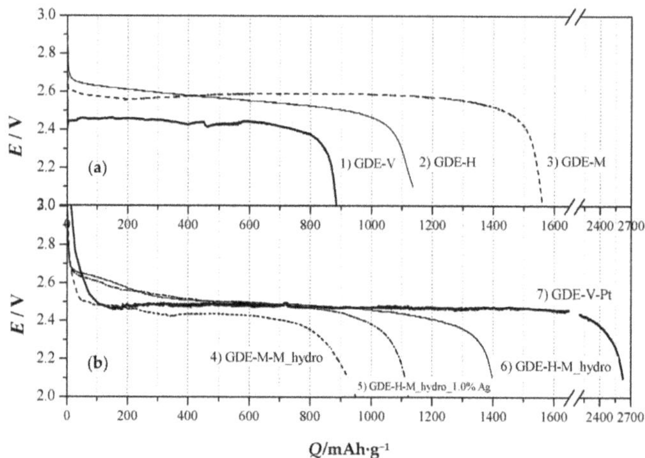

Figure 5. *E vs. Q* curves of GDEs by varying the (**a**) mesoporous carbon matrices and (**b**) carbonaceous supports added with MnO_2, during the first discharge cycle carried out at 0.34 mA·cm^{-2} in the Swagelok™-type cell (*S*-cell). Bold line: reference GDEs (GDE-V, Curve 1 and GDE-V-Pt, Curve 7).

Thus, the pores become quickly clogged by Li-oxides, shielding the surface suitable for crystals' nucleation. On the contrary, MCC, with larger pores (310 Å wide), easily accommodates the discharge products, presenting a larger surface on which nucleation can occur. This has a direct effect on the charge quantity effectively obtainable, thus proving that pore dimensions affect the specific capacity of the battery much more than the specific active area [27]. The performances of Li/air battery cathodes prepared with MnO_2 electrocatalytic powder added to a mesoporous carbon support are presented in Figure 5b, in which the reference cathode, GDE-V-Pt (Curve 7), is also included. The latter one shows the highest charge density (~2600 mAh·g^{-1}) during the first discharge cycle and an efficient discharge potential of about 2.5 V. This extremely high efficiency is due to the ORR (oxygen reduction reaction) electrocatalytic behavior of Pt powder supported on Vulcan, which affects also the plateau discharge potential; the value of 2.5 V *vs.* Li^+/Li is comparable to that already reported in the literature [49]. However, Pt is known to be a metal that prevents the further reaction of the discharge products (Li_2O and Li_2O_2), thus inhibiting the cyclability of the entire cell. Actually, in the case of GDE-V-Pt, a subsequent deactivation together with a dramatic decrease of the specific capacity after the first cycle occurs. Instead, the addition of the electrocatalytic M_hydro to the mesoporous carbon support reduces significantly the specific capacity furnished by MCC-based cathodes. Indeed, comparing GDE-M (Figure 5a, curve 3) and GDE-M-M_hydro (Figure 5b, curve 4), a loss of about 600 mAh·g^{-1} of the specific capacity can be noticed. This effect is probably ascribed to the pore size of the mesoporous support, which can be partially clogged by the MnO_2 powders, hence reducing the active surface area and concomitantly favoring the accommodation of the products of discharge cycles. This phenomenon can also explain the lower discharge potential of GDE-M-M_hydro when compared to GDE-V-Pt (2.4 *vs.* 2.5 V). On the contrary, M_hydro added to HCMSC based cathodes slightly increases the specific capacity of the composite electrodes; actually, comparing GDE-H (Figure 5a, Curve 2) and GDE-H-M_hydro (Figure 5b, Curve 6), an increase of more than 200 mAh·g^{-1} can be observed. Probably, the presence of mesoporosity in the MnO_2 samples, characterized by an order of magnitude higher than HCMSC distribution, does not allow any occlusion processes. Moreover, the larger surface area, the morphology of HCMSC and the homogeneous distribution of MnO_2 allow a better electron transfer process, thus making this GDE comparable to GDE-V-Pt, in terms of discharge potentials (2.5 V). It is worth noting that the addition of MnO_2 may catalyze the removal of LiO_2

and Li_2O_2 from the cathode surface during the charge cycles, thus increasing the durability and the cathode performances.

Finally, by analyzing the electrocatalytic performances of GDE-H-M_hydro_1.0%Ag (Figure 5b, Curve 5) and GDE-H (Figure 5a, Curve 2), the former seems not to extremely affect the HCMSC specific capacity, even if the Q value is 300 mAh· g^{-1} lower than that in the case of pure MnO_2 nanopowders.

For cyclability tests, an H-cell with a large volume has been employed in order to avoid electrolyte limitation due to its possible degradation [28–30]. In Figure 6, the specific capacity of GDE-H-M_hydro and GDE-H-M_hydro_1.0%Ag during the first five discharge cycles is reported. According to the literature [50–53], the first cycles contain the key information for a rapid screening of the electrode materials in terms of cell performances. Indeed, the chemical and morphological modifications of various composite electrodes [50–52] are comparable, since the limitations connected with the slow kinetics of the oxygen reactions [53] could be overcome. After the fifth cycle, the specific capacities of GDEs, prepared using all of the synthesized nanopowders, remain constant for the other five cycles. Unfortunately, when all of the air cathodes were stressed after ten cycles, the discharge capacity fades.

GDE-H-M_hydro, prepared using HCMSC carbon support with pure MnO_2, shows a dramatic decrease (around 50%) in specific capacity after the second run, highlighting the incomplete recharge processes by cycling consecutively. Instead, GDE-H-M_hydro_1.0%Ag shows a specific capacity up to 1400 mAh· g^{-1}, remaining above 1100 mAh· g^{-1} for all of the following cycles. Generally, the increase of specific capacity between the first and the second cycle for both cathodes can be ascribable to activation processes that allow a better accommodation and an easier removal of the solid discharge products. In particular, this effect is more evident for GDE-H-M_hydro_1.0%Ag (Figure 6, grey histograms) due to the presence of Ag dopant in the electrocatalytic powder, which plays an active role for the battery charge/discharge processes. Indeed, the addition of silver species leads to larger electrochemically active sites of the working material [54,55] and to transporting the insulating Li oxide products away from the carbon surface, in order to prevent it from blocking the electron transfer needed to reduce the O_2 [22]. The same behavior is underlined by the discharge/charge curves (E vs. Q, Figure 7) carried out at 0.34 mA· cm^{-2}.

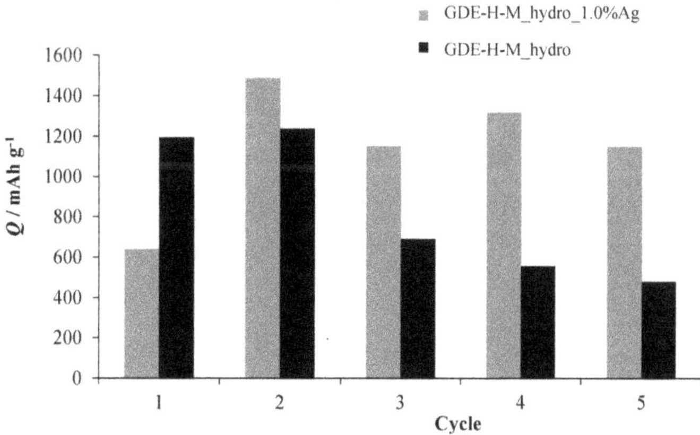

Figure 6. Specific capacity of GDE-H-M_hydro and GDE-H-M_hydro_1.0%Ag during the first five cycles in the home-made cell (H-cell) (black and grey histograms, respectively).

While GDE-H-M_hydro (Figure 7a) shows a rapid decrease of the specific capacity by cycling, in GDE-H-M_hydro_1.0%Ag (Figure 7b), a possible synergistic effect of MnO_2 and Ag occurs. In the present case, a stabilization of the discharge and charge curves above 1100 mAh· g^{-1} (from the third cycle on) for GDE-H-M_hydro_1.0%Ag is evident, as already seen in Figure 6. Moreover, the cut-off charge potential value for GDE-H-M_hydro is 4.3 V; at this potential, all of the tested cathodes undergo a rapid decrease of the performances, probably due to the presence of undesired discharge products. The other way around, the cut-off charge potential up to 4.5 V without any evidence of cathode degradation is noteworthy for GDE-H-M_hydro_1.0%Ag, pointing to the capability of Ag-doped MnO_2 electrocatalyst to easily remove the solid discharge products from the GDE pore structure.

These outcomes can be also appreciated looking at Figure 8a,b, in which the X-ray diffraction lines of GDE-H-M_hydro and GDE-H-M_hydro_1.0%Ag (obtained by subtraction of the respective patterns before and after five cycles) are reported. In all cases, the signals due to Li_2O_2 are almost absent for both GDEs; however, the presence of Li_2O (ICDD PDF-2 Card No. 077-2144) for both GDEs and of Li_2CO_3 (ICDD PDF-2 Card No. 022-1141), particularly for GDE-H-M_hydro (Figure 8a), is appreciable. Bruce *et al.* [26] discussed both the absence of Li_2O_2 and the presence of insoluble Li_2CO_3 coming from the degradation of carbonate solvents. In our case, the low amount of Li_2CO_3 for GDE-H-M_hydro_1.0%Ag justifies the less oxidative degradation of carbonate solvent and an easier removal of the discharge products from the GDE surface, probably due to the increased electronic conductivity of Ag-doped MnO_2 [22]. Actually, the presence of electrocatalytic material in GDE could help to reduce the solvent degradation [34], thus evidencing the possibility of a further increase of the solvent stability. Nowadays, the electrochemical stability of these compounds must be better investigated for their use in Li/Air devices, especially in the presence of suitable additives (vinylene carbonate, ethylene sulfite, *etc.*) able to stabilize the solvent [33], limiting degradation phenomena.

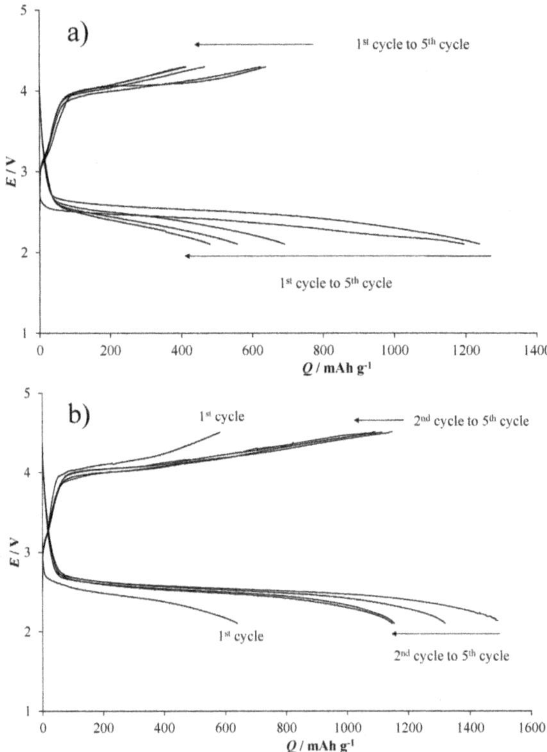

Figure 7. Discharge and charge cycles for (**a**) GDE-H-M_hydro and (**b**) GDE-H-M_hydro_1.0%Ag in the *H*-cell.

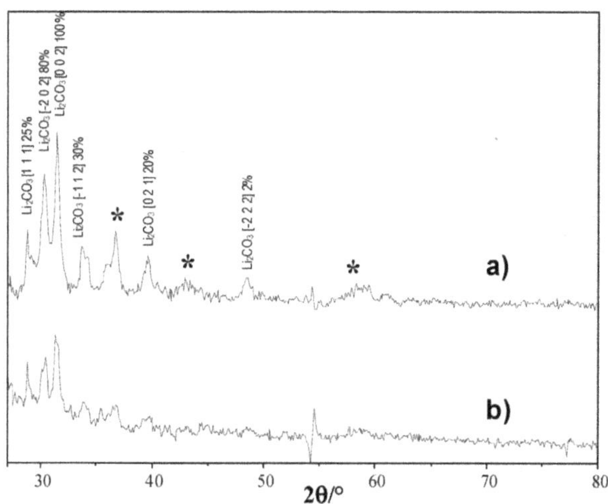

Figure 8. X-ray diffraction lines of (**a**) GDE-H-M_hydro and (**b**) GDE-H-M_hydro_1.0%Ag obtained by subtraction between the GDEs after five cycles and before cycling. The * label indicates the presence of MnO_2 polymorphs.

3. Experimental Section

3.1. Cathode Material and Gas Diffusion Electrode

In the present paper, specific attention has been given to cathodic materials. A general criterion in preparing GDEs was kept throughout all of the experiments: 20% by weight of manganese dioxide was added to a mixture of carbonaceous matrix (see the following). MnO_2 samples (see Table 1) were prepared via hydrothermal synthesis, operating a redox reaction of manganese sulfate with potassium permanganate (1:1 molar ratio) in 15 mL of Milli-Q (Darmstadt, Germany) water; reactants were mixed together and kept under stirring conditions at 90 °C for 24 h [38,56]. As a result, black powders were obtained, which were filtered, rinsed with Milli-Q water several times and finally dried at 60 °C for 24 h to provide the complete drying of MnO_2 (labeled as M_hydro in the following). A quote of pure MnO_2 was further subjected to a thermal treatment after the hydrothermal synthesis in the range 200–500 °C for 5 h in N_2 flux (named M_200 → M_500). A third group of samples was prepared to overcome the resistive behavior of manganese dioxide by adding $AgNO_3$ to the reaction mixture during the hydrothermal growth by varying the Ag/Mn molar ratio from 0.5% to 2% (called M_hydro_xAg, where x = molar ratio %). Zhang *et al.* [57] demonstrated that the increase in the conductivity leading to higher specific capacitance is due to the addition of silver into MnO_2 matrix.

Carbonaceous supports, a mesocellular carbon (MCC) and a hollow core mesoporous carbon (HCMSC), were prepared as widely described in the literature [46,58–60]. MCC shows a BET area of 231 $m^2 \cdot g^{-1}$ and a pore size of 310 Å, while HCMSC has a higher BET area, 1150 $m^2 \cdot g^{-1}$, with pores one order of magnitude smaller (38 Å). Empty HCMSC core shell particles create an extremely regular pattern of 400–500 nm diameter spherules.

GDEs (see Table 2) were prepared by mixing in N-methyl-2-pyrrolidinone (NMP 99.5%, anhydrous used as solvent, Sigma-Aldrich®, Milano, Italy), the mesoporous carbon and the manganese dioxide together with a small percentage by weight of Shawinigan Black AB50 carbon (SAB, Chevron Phillips, The Woodlands, TX, USA) and polyvinylidene fluoride (PVDF Kynar®, Arkema, Barcelona, Spain) powder, employed as a binder. The relative weight ratio among the components was: (MCC or HCMSC):SAB:PVDF = 65:20:15 or (MCC or HCMSC):MnO_2:SAB:PVDF = 45:20:20:15. After a 3-h stirring period, the slurry was brush-deposited, at 55 °C, onto a carbon paper sheet (SIGRACET®, SGL

Group, Meitingen, Germany). The complete evaporation of NMP was obtained by leaving the GDEs in an oven at 60 °C for 12 h. The electrode was punched into a 1.17-cm^2 disk with 1.3–2.2 mg· cm^{-2} of active material (excluding PVDF, the binder). Vulcan XC72R and Vulcan XC72R with 10% Pt were used to prepare GDEs to be tested as reference cathodes for the involved reactions. The catalytic activity and the physical characteristics of the carbon materials strongly affect the performance and the life of the electrodes. Maja *et al.* [61] reported that the proper selection of carbons leads to a significant enhancement of the performance of gas diffusion electrodes. Particularly, at long operation times, cathodes with Shawinigan acetylene black show better performances than those with oil-furnace carbon supports. Furthermore, the combination of Vulcan XC72R and SAB gives the best catalytic performances.

3.2. Cell Configuration

A solution consisting of 1 M $LiClO_4$ (Sigma-Aldrich®, battery grade, dry, ⩾99.99%), dissolved in propylene carbonate (PC: Sigma-Aldrich®, anhydrous, 99.7%), was used as the electrolyte. In an argon-filled (Gastec Vesta 5.5) glove box (LABCONCO®, Kansas, MO, USA), Li-air Swagelok™-type cells [62] were assembled stacking a Li disk (Sigma- Aldrich®, 99.9%) as the counter electrode, a glass fiber separator wetted with 90–100 μL of electrolyte and the gas diffusion working electrode. Concerning the choice of lithium salt, although it has no industrial application due to severe safety issues, $LiClO_4$ has been used in several academic studies reporting the high rechargeability of Li-O_2 cells. The relatively higher stability of $LiClO_4$ suggests it continues to be suitable for the purpose of tests and development at the research level [63]. Two different experimental cells were investigated: the first one, a Swagelok™ cell (S-cell) that ensured a rapid assembly, a handy configuration and an easy maintenance; the other one, a home-made cell (*H*-cell), in which the prepared GDEs, used as the cathode, can directly face a massive electrolyte solution (40 mL), in order to avoid the possible electrolyte limitations [27]. The anode was also employed as a reference electrode, and potential values have been always referred to the Li^+/Li redox couple. Oxygen (Gastec Vesta 5.5) was fed to the gas side of the GDEs at a 1-mL· s^{-1} flow, during both the charging and discharging period. Both of the cells were studied performing chronopotentiometric experiments (galvanostatic analysis) imposing a discharge and charge current density and cut-off potential of 0.34 mA· cm^{-2} and of 4.3 V (*vs.* Li/Li^+), respectively.

3.3. Instrumentation

A Princeton Applied Research 263A potentiostat/galvanostat was employed to perform cycle tests, and CorrWare 3.1c software (Oak Bridge, TN, USA) was used for data acquisition. Room-temperature X-ray powder diffraction (XRPD) patterns were collected between 20 and 90° ($\Delta 2\theta$ = 0.08°, time per step = 17 s, scan speed = 0.02°/s) with a Siemens D500 diffractometer (Berlin, Germany), using Cu K_α radiation (Cu $K_{\alpha 1}$: λ = 1.54056 Å, $K_{\alpha 2}$: λ = 1.54433 Å). Specific surface areas were determined by the classical BET procedure using a Coulter SA 3100 apparatus (BeckmanCoulter, Fullerton, CA, USA). Desorption isotherms were used to determine the pore size distribution using the Barret Joyner Halenda (BJH) method. Particle morphology was obtained by scanning electron microscopy (SEM), using a Hitachi TM-1000 (Tokyo, Japan) equipped with an energy dispersive X-ray (EDS, SwiftED-TM) detector. Transmission electron microscope and scanning transmission electron microscope (TEM/STEM) analyses were performed by a ZEISS LIBRA200 EFTEM instrument (Jena, Germany) operated at 200 kV. The STEM micrographs were collected by a high-angle annular dark field (HAADF) detector. The TEM specimens were prepared suspending the nanoparticles in isopropanol, sonicating the suspension in an ultrasonic bath for 20 min and dropping it onto a holey carbon-coated Cu grid. The specimens were analyzed after complete solvent evaporation in air at room temperature.

4. Conclusions

Lithium-air batteries represent valid devices in facing the increasingly worldwide energy demand. Nevertheless, before they can be fully developed and employed in different fields, from automotive to portable devices, cathode optimization is requested together with the development of a new family of low cost electrolytes, able to resist the presence of oxygen radicals. Nowadays, the optimization of cathodic materials is required to develop batteries able to produce high specific energy.

In this work GDE composites, made of two diverse mesoporous carbons (MCC and HCMSC) and bare and Ag-doped MnO_2 electrocatalysts, have been deeply characterized and electrochemically tested in home-made Li/air cells. All hydrothermally-synthesized MnO_2 nanoparticles show a nanoneedle structure; particularly, M_hydro presents the best properties in terms of surface area and total pore volume distribution. As concerns the doped samples, only M_hydro_1.0%Ag has been tested due to its homogeneous distribution in the MnO_2 matrix.

The behavior of the GDE obtained by using only mesoporous carbons depends on their pore size distribution: MCC with larger pores (310 Å) shows a specific capacity of 1560 $mAh \cdot g^{-1}$, while HCMSC (d = 40 Å) reaches a value around 1150 $mAh \cdot g^{-1}$. When electrocatalytic M_hydro nanopowders are incorporated into the composite GDE, a drastic decrease of performance can be evidenced for MCC-based cathodes, whereas no effects are significant in the case of HCMSC-based ones. Hydrothermal MnO_2 doped with Ag 1.0% shows an average specific capacity above 1100 $mAh \cdot g^{-1}$ during cycles. This effect can be due to silver atoms in a manganese oxide lattice that induce an electrical conductivity improvement and limit the presence of Li-based discharge products. The combination of Ag with MnO_2 could lead to composite cathodes for Li/air batteries that present high specific capacity, thus reducing the overpotentials and increasing the cyclability of the cell.

Supplementary Materials: They are available online at http://www.mdpi.com/2079-4991/6/1/10/s1.

Acknowledgments: This research has been supported by Fondazione Cariplo (2010-0506). Nerino Penazzi, Silvia Bodoardo and Carlotta Francia, Politecnico di Torino, are gratefully acknowledged for the preparation and characterization of mesoporous carbon matrices (HCMSC and MCC).

Author Contributions: A.M., G.L., C.L. S.R. and A.V.: fabrication of GDEs, *H*-cell design and assembly, electrochemical characterizations of GDEs, definition of electrochemical test to verify cathodes performances and data elaboration. G.C., E.P., C.D.B. and M.M.: synthesis of bare and doped electrocatalytic nanopowders and physico-chemical characterizations.

Conflicts of Interest: The authors declare that they have no conflict of interest.

References

1. Abraham, K.M.; Jiang, Z. A Polymer Electrolyte-Based Rechargeable Lithium/Oxygen Battery. *J. Electrochem. Soc.* **1996**, *143*, 1–5. [CrossRef]
2. Girishkumar, G.; McCloskey, B.; Luntz, A.C.; Swanson, S.; Wilcke, W. Lithium-air battery: Promise and challenges. *J. Phys. Chem. Lett.* **2010**, *1*, 2193–2203. [CrossRef]
3. Minguzzi, A.; Locatelli, C.; Cappelletti, G.; Bianchi, C.L.; Vertova, A.; Ardizzone, S.; Rondinini, S. Designing materials by means of the cavity-microelectrode: The introduction of the quantitative rapid screening toward a highly efficient catalyst for water oxidation. *J. Mater. Chem.* **2012**, *22*, 8896–8902. [CrossRef]
4. Minguzzi, A.; Locatelli, C.; Lugaresi, O.; Vertova, A.; Rondinini, S. Au-based/electrochemically etched cavity-microelectrodes as optimal tool for quantitative analyses on finely dispersed electrode materials: Pt/C, IrO_2-SnO_2 and Ag catalysts. *Electrochim. Acta* **2013**, *114*, 637–642. [CrossRef]
5. Locatelli, C.; Minguzzi, A.; Vertova, A.; Cava, P.; Rondinini, S. Quantitative studies on electrode material properties by means of the cavity microelectrode. *Anal. Chem.* **2011**, *83*, 2819–2823. [CrossRef] [PubMed]
6. Minguzzi, A.; Lugaresi, O.; Locatelli, C.; Rondinini, S.; D'Acapito, F.; Achilli, E.; Ghigna, P. Fixed energy X-ray absorption voltammetry. *Anal. Chem.* **2013**, *85*, 7009–7013. [CrossRef] [PubMed]
7. Zhang, S.S.; Foster, D.; Read, J. Discharge characteristic of a non-aqueous electrolyte Li/O_2 battery. *J. Power Sources* **2010**, *195*, 1235–1240. [CrossRef]

8. Lu, Y.-C.; Gasteiger, H.A.; Crumlin, E.; McGuire, R.; Shao-Horn, Y. Electrocatalytic Activity Studies of Select Metal Surfaces and Implications in Li-Air Batteries. *J. Electrochem. Soc.* **2010**, *157*, A1016–A1025. [CrossRef]
9. Hassoun, J.; Croce, F.; Armand, M.; Scrosati, B. Investigation of the O_2 electrochemistry in a polymer electrolyte solid-state cell. *Angew. Chem. Int. Ed.* **2011**, *50*, 2999–3002. [CrossRef] [PubMed]
10. Padbury, R.; Zhang, X. Lithium–oxygen batteries—Limiting factors that affect performance. *J. Power Sources* **2011**, *196*, 4436–4444. [CrossRef]
11. Rahman, M.A.; Wang, X.; Wen, C. A review of high energy density lithium-air battery technology. *J. Appl. Electrochem.* **2014**, *44*, 5–22. [CrossRef]
12. Cheng, H.; Scott, K. Nano-structured gas diffusion electrode—A high power and stable cathode material for rechargeable Li-air batteries. *J. Power Sources* **2013**, *235*, 226–233. [CrossRef]
13. Calegaro, M.L.; Lima, F.H.B.; Ticianelli, E.A. Oxygen reduction reaction on nanosized manganese oxide particles dispersed on carbon in alkaline solutions. *J. Power Sources* **2006**, *158*, 735–739. [CrossRef]
14. Ma, Y.; Wang, H.; Ji, S.; Goh, J.; Feng, H.; Wang, R. Highly active Vulcan carbon composite for oxygen reduction reaction in alkaline medium. *Electrochim. Acta* **2014**, *133*, 391–398. [CrossRef]
15. Tran, C.; Yang, X.Q.; Qu, D. Investigation of the gas-diffusion-electrode used as lithium/air cathode in non-aqueous electrolyte and the importance of carbon material porosity. *J. Power Sources* **2010**, *195*, 2057–2063. [CrossRef]
16. Xiao, J.; Wang, D.; Xu, W.; Wang, D.; Williford, R.E.; Liu, J.; Zhang, J.-G. Optimization of Air Electrode for Li/Air Batteries. *J. Electrochem. Soc.* **2010**, *157*, A487–A492. [CrossRef]
17. Débart, A.; Bao, J.; Armstrong, G.; Bruce, P.G. An O_2 cathode for rechargeable lithium batteries: The effect of a catalyst. *J. Power Sources* **2007**, *174*, 1177–1182. [CrossRef]
18. Locatelli, C.; Minguzzi, A.; Vertova, A.; Rondinini, S. IrO_2–SnO_2 mixtures as electrocatalysts for the oxygen reduction reaction in alkaline media. *J. Appl. Electrochem.* **2013**, *43*, 171–179. [CrossRef]
19. Minguzzi, A.; Locatelli, C.; Cappelletti, G.; Scavini, M.; Vertova, A.; Ghigna, P.; Sandra Rondinini, S. IrO_2-based disperse-phase electrocatalysts: A complementary study by means of the cavity-microelectrode and ex-situ X-ray absorption spectroscopy. *J. Phys. Chem. A* **2012**, *116*, 6497–6504. [CrossRef] [PubMed]
20. Chen, Y.; Freunberger, S.A.; Peng, Z.; Fontaine, O.; Bruce, P.G. Charging a Li-O_2 battery using a redox mediator. *Nat. Chem.* **2013**, *5*, 489–494. [CrossRef] [PubMed]
21. Lim, H.D.; Song, H.; Kim, J.; Gwon, H.; Bae, Y.; Park, K.Y.; Hong, J.; Kim, H.; Kim, T.; Kim, Y.H.; *et al.* Superior rechargeability and efficiency of lithium-oxygen batteries: Hierarchical air electrode architecture combined with a soluble catalyst. *Angew. Chem. Int. Ed.* **2014**, *53*, 3926–3931. [CrossRef] [PubMed]
22. Sun, D.; Shen, Y.; Zhang, W.; Yu, L.; Yi, Z.; Yin, W.; Wang, D.; Huang, Y.; Wang, J.; Wang, D.; *et al.* A solution-phase bifunctional catalyst for lithium-oxygen batteries. *J. Am. Chem. Soc.* **2014**, *136*, 8941–8946. [CrossRef] [PubMed]
23. Cheng, H.; Scott, K. Selection of oxygen reduction catalysts for rechargeable lithium-air batteries-metal or oxide? *Appl. Catal. B* **2011**, *108–109*, 140–151. [CrossRef]
24. Kalubarme, R.S.; Cho, M.-S.; Yun, K.-S.; Kim, T.-S.; Park, C.-J. Catalytic characteristics of MnO_2 nanostructures for the O_2 reduction process. *Nanotechnology* **2011**, *22*. [CrossRef] [PubMed]
25. Xu, W.; Viswanathan, V.V.; Wang, D.; Towne, S.A.; Xiao, J.; Nie, Z.; Hu, D.; Zhang, J.G. Investigation on the charging process of Li_2O_2-based air electrodes in Li-O_2 batteries with organic carbonate electrolytes. *J. Power Sources* **2011**, *196*, 3894–3899. [CrossRef]
26. Freunberger, S.A.; Chen, Y.; Peng, Z.; Griffin, J.M.; Hardwick, L.J.; Bardé, F.; Novák, P.; Bruce, P.G. Reactions in the rechargeable lithium-O_2 battery with alkyl carbonate electrolytes. *J. Am. Chem. Soc.* **2011**, *133*, 8040–8047. [CrossRef] [PubMed]
27. Shao, Y.; Ding, F.; Xiao, J.; Zhang, J.; Xu, W.; Park, S.; Zhang, J.G.; Wang, Y.; Liu, J. Making Li-air batteries rechargeable: Material challenges. *Adv. Funct. Mater.* **2013**, *23*, 987–1004. [CrossRef]
28. Bardenhagen, I.; Fenske, M.; Fenske, D.; Wittstock, A.; Bäumer, M. Distribution of discharge products inside of the lithium/oxygen battery cathode. *J. Power Sources* **2015**, *299*, 162–169. [CrossRef]
29. Bardenhagen, I.; Yezerska, O.; Augustin, M.; Fenske, D.; Wittstock, A.; Bäumer, M. *In situ* investigation of pore clogging during discharge of a Li/O_2 battery by electrochemical impedance spectroscopy. *J. Power Sources* **2015**, *278*, 255–264. [CrossRef]

30. Marchini, F.; Herrera, S.; Torres, W.; Tesio, A.Y.; Williams, F.J.; Calvo, E.J. Surface Study of Lithium–Air Battery Oxygen Cathodes in Different Solvent–Electrolyte Pairs. *Langmuir* **2015**, *31*, 9236–9245. [CrossRef] [PubMed]
31. Marinaro, M.; Balasubramanian, P.; Gucciardi, E.; Theil, S.; Jçrissen, L. Importance of Reaction Kinetics and Oxygen Crossover in aprotic Li-O_2 Batteries Based on a Dimethyl Sulfoxide Electrolyte. *ChemSusChem* **2015**, *8*, 3139–3145. [CrossRef] [PubMed]
32. Kim, D.S.; Park, Y.J. Effect of multi-catalysts on rechargeable Li-air batteries. *J. Alloys Compd.* **2014**, *591*, 164–169. [CrossRef]
33. Ogasawara, T.; Débart, A.; Holzapfel, M.; Novák, P.; Bruce, P.G. Rechargeable Li_2O_2 electrode for lithium batteries. *J. Am. Chem. Soc.* **2006**, *128*, 1390–1393. [CrossRef] [PubMed]
34. Débart, A.; Paterson, A.J.; Bao, J.; Bruce, P.G. α-MnO_2 nanowires: A catalyst for the O_2 electrode in rechargeable lithium batteries. *Angew. Chem. Int. Ed.* **2008**, *47*, 4521–4524. [CrossRef] [PubMed]
35. Turner, S.; Buseck, P.R. Defects in nsutite (γ-MnO_2) and dry-cell battery efficiency. *Nature* **1983**, *304*, 143–146. [CrossRef]
36. De Wolff, P.M. Interpretation of some γ-MnO_2 diffraction patterns. *Acta Crystallogr.* **1959**, *12*, 341–345. [CrossRef]
37. Chabre, Y.; Pannetier, J. Structural and electrochemical properties of the proton/γ-MnO_2 system. *Prog. Solid State Chem.* **1995**, *23*, 1–130. [CrossRef]
38. Benhaddad, L.; Makhloufi, L.; Messaoudi, B.; Rahmouni, K.; Takenouti, H. Reactivity of Nanostructured MnO_2 in Alkaline Medium Studied with a Micro-Cavity Electrode: Effect of Synthesizing Temperature. *ACS Appl. Mater. Interfaces* **2009**, *1*, 424–432. [CrossRef] [PubMed]
39. Muraoka, Y.; Chiba, H.; Atou, T.; Kikuchi, M.; Hiraga, K.; Syono, Y. Preparation of γ-MnO_2 with an Open Tunnel. *J. Solid State Chem.* **1999**, *144*, 136–142. [CrossRef]
40. Ching, S.; Petrovay, D.J.; Jorgensen, M.L.; Suib, S.L. Sol−Gel Synthesis of Layered Birnessite-Type Manganese Oxides. *Inorg. Chem.* **1997**, *36*, 883–890. [CrossRef]
41. Reddy, R.N.; Reddy, R.G. Sol-gel MnO_2 as an electrode material for electrochemical capacitors. *J. Power Sources* **2003**, *124*, 330–337. [CrossRef]
42. Byström, A.M.; Lund, E.W.; Lund, L.K.; Hakala, M. The Crystal Structure of Ramsdellite, an Orthorhombic Modification of MnO_2. *Acta Chem. Scand.* **1949**, *3*, 163–173. [CrossRef]
43. Baur, W.H. Rutile-type compounds. V. Refinement of MnO_2 and MgF_2. *Acta Crystallogr. Sect. B* **1976**, *32*, 2200–2204. [CrossRef]
44. Post, J.E.; von Dreele, R.B.; Buseck, P.R. Symmetry and cation displacements in hollandites: Structure refinements of hollandite, cryptomelane and priderite. *Acta Crystallogr. Sect. B* **1982**, *38*, 1056–1065. [CrossRef]
45. Geller, S. Structure of α-Mn_2O_3, $(Mn_{0.983}Fe_{0.017})_2O_3$ and $(Mn_{0.37}Fe_{0.63})_2O_3$ and relation to magnetic ordering. *Acta Crystallogr. Sect. B* **1971**, *27*, 821–828. [CrossRef]
46. Joo, J.B.; Kim, P.; Kim, W.; Kim, J.; Yi, J. Preparation of mesoporous carbon templated by silica particles for use as a catalyst support in polymer electrolyte membrane fuel cells. *Catal. Today* **2006**, *111*, 171–175. [CrossRef]
47. Wang, F.; Xu, Y.-H.; Luo, Z.-K.; Pang, Y.; Wu, Q.-X.; Liang, C.-S.; Chen, J.; Liu, D.; Zhang, X. A dual pore carbon aerogel based air cathode for a highly rechargeable lithium-air battery. *J. Power Sources* **2014**, *272*, 1061–1071. [CrossRef]
48. Cetinkaya, T.; Ozcan, S.; Uysal, M.; Guler, M.O.; Akbulut, H. Free-standing flexible graphene oxide paper electrode for rechargeable Li-O_2 batteries. *J. Power Sources* **2014**, *267*, 140–147. [CrossRef]
49. Lu, Y.-C.; Gasteiger, H.A.; Parent, M.C.; Chiloyan, V.; Shao-Horn, Y. The Influence of Catalysts on Discharge and Charge Voltages of Rechargeable Li-Oxygen Batteries. *Electrochem. Solid-State Lett.* **2010**, *13*, A69–A72. [CrossRef]
50. Du, Z.; Yang, P.; Wang, L.; Lu, Y.; Goodenough, J.B.; Zhang, J.; Zhang, D. Electrocatalytic performances of $LaNi_{1-x}Mg_xO_3$ perovskite oxides as bi-functional catalysts for lithium air batteries. *J. Power Sources* **2014**, *265*, 91–96. [CrossRef]
51. Zhang, L.; Zhang, F.; Huang, G.; Wang, J.; Du, X.; Qin, Y.; Wang, L. Freestanding MnO_2@carbon papers air electrodes for rechargeable Li-O_2 batteries. *J. Power Sources* **2014**, *261*, 311–316. [CrossRef]
52. Hsu, C.-H.; Shen, Y.-W.; Chien, L.-H.; Kuo, P.-L. Li_2FeSiO_4 nanorod as high stability electrode for lithium-ion batteries. *J. Nanoparticle Res.* **2015**, *17*, 1–9. [CrossRef]

53. Suntivich, J.; Gasteiger, H.A.; Yabuuchi, N.; Shao-Horn, Y. Electrocatalytic Measurement Methodology of Oxide Catalysts Using a Thin-Film Rotating Disk Electrode. *J. Electrochem. Soc.* **2010**, *157*, B1263–B1268. [CrossRef]
54. Ardizzone, S.; Bianchi, C.L.; Cappelletti, G.; Ionita, M.; Minguzzi, A.; Rondinini, S.; Vertova, A. Composite ternary SnO_2-IrO_2-Ta_2O_5 oxide electrocatalysts. *J. Electroanal. Chem.* **2006**, *589*, 160–166. [CrossRef]
55. Ionita, M.; Cappelletti, G.; Minguzzi, A.; Ardizzone, S.; Bianchi, C.; Rondinini, S.; Vertova, A. Bulk, Surface and Morphological Features of Nanostructured Tin Oxide by a Controlled Alkoxide-Gel Path. *J. Nanoparticle Res.* **2006**, *8*, 653–660. [CrossRef]
56. Benhaddad, L.; Bazin, C.; Makhloufi, L.; Messaoudi, B.; Pillier, F.; Rahmouni, K.; Takenouti, H. Effect of synthesis duration on the morphological and structural modification of the sea urchin-nanostructured γ-MnO_2 and study of its electrochemical reactivity in alkaline medium. *J. Solid State Electrochem.* **2014**, *18*, 2111–2121. [CrossRef]
57. Zhang, G.; Zheng, L.; Zhang, M.; Guo, S.; Liu, Z.H.; Yang, Z.; Wang, Z. Preparation of Ag-nanoparticle-loaded MnO_2 nanosheets and their capacitance behavior. *Energy Fuels* **2012**, *26*, 618–623. [CrossRef]
58. Zeng, J.; Francia, C.; Dumitrescu, M.A.; Monteverde Videla, A.H.A.; Ijeri, V.S.; Specchia, S.; Spinelli, P. Electrochemical performance of Pt-based catalysts supported on different ordered mesoporous carbons (Pt/OMCs) for oxygen reduction reaction. *Ind. Eng. Chem. Res.* **2012**, *51*, 7500–7509. [CrossRef]
59. Büchel, G.; Unger, K.K.; Matsumoto, A.; Tsutsumi, K. A Novel Pathway for Synthesis of Submicrometer-Size Solid Core/Mesoporous Shell Silica Spheres. *Adv. Mater.* **1998**, *10*, 1036–1038. [CrossRef]
60. Zeng, J.; Francia, C.; Gerbaldi, C.; Dumitrescu, M.A.; Specchia, S.; Spinelli, P. Smart synthesis of hollow core mesoporous shell carbons (HCMSC) as effective catalyst supports for methanol oxidation and oxygen reduction reactions. *J. Solid State Electrochem.* **2012**, *16*, 3087–3096. [CrossRef]
61. Maja, M.; Orecchia, C.; Strano, M.; Tosco, P.; Vanni, M. Effect of structure of the electrical performance of gas diffusion electrodes for metal air batteries. *Electrochim. Acta* **2000**, *46*, 423–432. [CrossRef]
62. Beattie, S.D.; Manolescu, D.M.; Blair, S.L. High-Capacity Lithium-Air Cathodes. *J. Electrochem. Soc.* **2009**, *156*, A44–A47. [CrossRef]
63. Younesi, R.; Veith, G.M.; Johansson, P.; Edström, K.; Vegge, T. Lithium salts for advanced lithium batteries: Li–metal, Li–O_2, and Li–S. *Energy Environ. Sci.* **2015**, *8*, 1905–1922. [CrossRef]

MDPI AG
St. Alban-Anlage 66
4052 Basel, Switzerland
Tel. +41 61 683 77 34
Fax +41 61 302 89 18
http://www.mdpi.com

Nanomaterials Editorial Office
E-mail: nanomaterials@mdpi.com
http://www.mdpi.com/journal/nanomaterials

www.ingramcontent.com/pod-product-compliance
Lightning Source LLC
Chambersburg PA
CBHW041218220326
41597CB00033BA/6011

* 9 7 8 3 0 3 8 4 2 5 3 6 6 *